LIBRARY
Astronomy and Astrophysics Library

The investigation of cosmic phenomena is one of mankind's oldest adventures and at the same time one of the areas where we are presently witnessing the most astounding expansion of knowledge in the natural sciences. The rapid growth of experimental evidence due to, e.g., modern computer technology, satellite-borne observatories and high-tech instrumentation even tends to accelerate this. To keep abreast, scientists collect their information in specialized journals, proceedings of international symposia and tersely written review articles. These sources are hardly accessible to students.

Textbooks at the advanced graduate level have to bridge the gap from introductory texts to original work published for the expert. A&A Library is designed to meet this aim as a series of high-quality textbooks and well-balanced (documented) research monographs for the advanced student. Books published in A&A Library present proven, mature facts and expose differing opinions on open problems and will guide the best students to the "hot" topics of today's research in astronomy and astrophysics.

The series A&A Library, however, is more. It aims to close the gap between the astronomical sciences and other areas of modern physics. It meets the growing need to critically examine the experimental and theoretical results and recognize both the lasting and the short-lived. Its broad coverage and its careful supervision by the editors make A&A Library a reliable source of information for researchers and students alike. As in any other field of science, teaching and training at the graduate level is invaluable for the advancement of astronomy and astrophysics. Addressing students at a point when they are specializing in a particular discipline, advanced classes are decisive in the quality of the next generation of researchers. A&A Library will provide them with modern textbooks and monographs to meet the most ambitious demands.

More information about this series at https://link.springer.com/bookseries/848

Debades Bandyopadhyay • Kamales Kar

Supernovae, Neutron Star Physics and Nucleosynthesis

 Springer

Debades Bandyopadhyay (iD)
Astroparticle Physics and Cosmology
Division
Saha Institute of Nuclear Physics
Kolkata, India

Kamales Kar (iD)
Department of Physics
RKMVERI
Belur Math, West Bengal, India

ISSN 0941-7834   ISSN 2196-9698   (electronic)
Astronomy and Astrophysics Library
ISBN 978-3-030-95173-3   ISBN 978-3-030-95171-9   (eBook)
https://doi.org/10.1007/978-3-030-95171-9

Cover illustration: creativemarc/Shutterstock.com

This Springer imprint is published by the registered company Springer Nature Switzerland AG
The registered company address is: Gewerbestrasse 11, 6330 Cham, Switzerland

To our parents

Foreword

This monograph, *Supernovae, Neutron Star Physics and Nucleosynthesis*, written by Debades Bandyopadhyay and Kamales Kar, is very timely. Neutron stars, especially binary neutron stars, have a central role in interpreting the recently discovered gravitational waves using the gravitational wave observatories LIGO and Virgo. The Facility for Antiproton and Ion Research, FAIR, at the GSI site of Germany is dedicated, among other commitments, to the study of hot dense matter, exactly the stuff that neutron stars are made of. India is involved in LIGO as well as FAIR using the CBM (Compressed Baryonic Matter) detector. This well-written monograph will have great relevance for them. The other mystery is to do with SN1987A. For a very long time, 34 years or more, astrophysicists have been trying to find out what exactly happened at the core after the supernova explosion. It remained a mystery, shrouded by an opaque curtain of dust. It is only now that the cloud of cosmic dust has settled down, we know for sure that at the core there is a compact object, most probably a neutron star. Since the discovery of the star in 1967 by Jocelyn Bell, Antony Hewish and Martin Ryle, the neutron star continues to be an enigma. Wonders are still coming up!

The core collapse of massive stars has been an outstanding problem and truly interdisciplinary, and this has been extensively covered in this book. Rather, more recently, binary neutron stars have turned out to be an indispensable laboratory for detection and study of gravitational waves. In fact, it is clear that binary neutron stars and black holes will continue to be the testing ground for the newly discovered field of gravitational waves and gravitational radiation.

Finally, a tribute goes to both the authors, formerly senior faculty members at Saha Institute of Nuclear Physics, for writing this monograph so lucidly. The book is a handsome salute to its founder, Meghnad Saha, who ushered in observational astrophysics way back in the early twentieth century by discovering the Saha Ionisation Equation.

For both the initiated as well as beginners of today's nuclear astrophysics and cosmology, this monograph is a must read.

INSA Senior Scientist   Bikash Sinha
Former Homi Bhabha Professor
Department of Atomic Energy
Former Director, Saha Institute of Nuclear Physics
and Variable Energy Cyclotron Centre
Department of Atomic Energy
Kolkata, India
November 2021

Preface

The second half of the last decade had been very eventful for astrophysics in particular and physics in general. Several outstanding astrophysical discoveries during this period have impacted physics research, cutting across different disciplines of physics. The most prominent among those was the first discovery of gravitational waves from two colliding neutron stars in the binary on 17 August 2017 and denoted as GW170817. That year was also the golden jubilee year of the discovery of the first pulsar by Jocelyn Bell and Antony Hewish. The first binary merger GW170817 was fascinating because it was observed across the electromagnetic spectrum. This laid the foundation of a new era in the multimessenger astronomy along with gravitation waves. The other astrophysical observation that has generated lots of curiosity is the possible detection of a compact object inside the supernova remnant of SN1987A. Explaining the physics of core collapse supernova explosions and the problem of production of heavy elements through nucleosynthesis involved sustained activity over the last 50 years. However, now, with the results of very detailed simulations available, one is able to get a clearer picture of the issues involved. The study of core collapse supernova explosion mechanism, the birth of neutron stars, and binary neutron star mergers is finely related to each other and is the central attraction of nuclear astrophysics. This is a very thriving interdisciplinary area of research. This will be well supported by celestial observations with the advent of the new generation of telescopes, such as James Webb Space Telescope, Square Kilometre Array, and gravitational wave observatory LIGO, India, and the terrestrial experiments, for example, the investigation of hot and dense matter at the Facility for Antiproton and Ion Research, Germany, in this decade. It is very timely to write a book on these topical issues in nuclear astrophysics.

The book assumes the readers to have knowledge of basic nuclear physics and some exposure to quantum field theory and general relativity. References to background material have been given adequately.

The encouragement of B. Ananthanarayan to write this book is highly appreciated. Further, we thank the Springer Nature team, in particular Lisa Scalone, for guiding us through the writing of this book.

Debades Bandyopadhyay (DB) humbly acknowledges the contributions of his collaborators particularly, S. K. Samaddar, J. N. De, S. Chakraborty, S. Pal, Sarmistha Banik, Debarati Chatterjee, Rana Nandi, Chandrachur Chakraborty, Apurb Kheto, Prasanta Char, Sajad Ahmad Bhat, Monika Sinha, Shriya Soma, Arunava Mukherjee, Matthias Hanauske, and Matthias Hempel. DB is grateful to Bikash Sinha and Horst Stöcker for their support and encouragement.

Kamales Kar acknowledges the help he received from Ramakrishna Mission Vivekananda Educational and Research Institute for the affiliation and support, and the encouragement from its ex-vice chancellor Swami Atmapriyananda. In addition, he thanks Abhijit Bandyopadhyay, Pijushpani Bhattacharjee, Soumya Chakravarti, Sandhya Choubey, Palash Baran Pal for discussions and M.V. Krishnan for encouragement. Kamales Kar is also indebted to his wife Chitra and daughter Aparajita for their enthusiasm and support at all stages of the writing of this book.

Kolkata, India   Debades Bandyopadhyay
Belur Math, India   Kamales Kar
November 2021

Contents

Acronyms

1D	One Dimensional
2SC	Two Flavor Color Superconducting
AAS	American Astronomical Society
AGB	Asymptotic Giant Branch
APR	Akmal, Pandharipande and Ravenhall
APS	American Physical Society
BBFH	Burbidge, Burbidge, Fowler and Hoyle
BBH	Binary Black Hole
BE	Binding Energy
BH	Black Hole
BHB	Banik, Hempel and Bandyopadhyay
BHF	Brueckner-Hartree-Fock
BBHNS	Binary Black Hole Neutron Star
BBP	Baym, Bethe and Pethick
BLV	Bonche, Levit and Vautherin
BNS	Binary Neutron Star
BPS	Baym, Pethick and Sutherland
BRUSLIB	Brussels Nuclear Library for Astrophysics Applications
BSG	Blue Supergiant
Cas A	Cassiopeia A
CC	Charged Current
CCSN	Core Collapse Supernova
CFL	Color-Flavor-Locked
ChEFT	Chiral Effective Field Theory
DB	Dirac-Brueckner
DD	Density Dependent
DBHF	Dirac-Brueckner Hartree-Fock
DDRH	Density Dependent Relativistic Hadron
DSNB	Diffuse Supernova Neutrino Background
EC	Electron Capture
EFT	Effective Field Theory

EM	Electromagnetic
EoS	Equation of State
ETFSI	Extended Thomas-Fermi with Strutinski Integral
FAIR	Facility for Antiproton and Ion Research
FRDM	Finite-Range Droplet Model
GRB	Gamma Ray Bursts
GR1D	General Relativistic One Dimensional
GM	Glendenning and Moszkowski
GT	Gamow-Teller
GWs	Gravitational Waves
HMNS	Hypermassive Neutron Star
HR	Hertzsprung-Russell
HS	Hempel-Schaffner-Bielich
HST	Hubble Space Telescope
IMB	Irvine-Michigan-Brookhaven
K II	Kamiokande II
KEH	Komatsu-Eriguchi-Hachisu
KN	Kilonova
kpc	Kiloparsec
LIGO	Laser Interferometer Gravitational-wave Observatory
LISA	Laser Interferometer Space Antenna
LMC	Large Magellanic Cloud
LVC	LIGO-Virgo collaboration
MES	Multi-Event s-process
MHD	Magnetohydrodynamic
MIT	Massachusetts Institute of Technology
MSP	Millisecond Pulsar
NEMO	Neutron Star Extreme Matter Observatory
NICER	Neutron Star Interior Composition Explorer
NG	Nucleus plus Gas
NSE	Nuclear Statistical Equilibrium
PDF	Posterior Probability Density
PM	Post-Newtonian
PNS	Protoneutron Star
PSR	Pulsar
PTA	Pulsar Timing Array
QCD	Quantum Chromodynamics
QHD	Quantum Hadrodynamics
QRPA	Quasi-particle Random Phase Approximation
RMF	Relativistic Mean Field
RNS	Rapidly Rotating Neutron Star
RPA	Random Phase Approximation
RSG	Red Supergiant
SDD	Scalar Density Dependence
SFHo	Steiner-Fischer-Hempel Optimal

SFHx	Steiner-Fischer-Hempel Extremal
sGRB	short Gamma Ray Burst
SK	Super Kamioka
SKA	Square Kilometre Array
SkM	Skyrme
SMMC	Shell Model Monte Carlo
SMNS	Supermassive Neutron Star
SN	Supernova
SNEWS	Supernova Early Warning System
SNO	Sudbury Neutrino Observatory
SO	Spin-Orbit
SoS	Solar System
TF	Thomas-Fermi
TOV	Tolman-Oppenheimer-Volkoff
TPC	Time Projection Chamber
VDD	Vector Density Dependence
WH	Woosley and Heger
WS	Wigner-Seitz

Chapter 1
Introduction

Nuclear astrophysics, the overlap area of nuclear physics and astrophysics, has over the years become one of the central areas of physics research, giving answers to challenging problems that were unsolved for a long time. This also has generated a big appeal to the imaginative mind that the laws and structure of the microscopic world of the smallest particles are playing crucial roles in the explanation of events of the macroscopic world involving the largest objects like huge stars. But some of these ideas and conjectures took time to get accepted by the physics community and follow-up works started after decades. For example, the supernovae were conjectured as connected to the collapse of stars in 1933–1934 and Subrahmanyan Chandrasekhar in 1931 showed that the pressure of the relativistic degenerate electron gas alone cannot balance the inward gravitational pressure in stars having core masses larger than a critical mass, leading to their collapse. But for reaching that stage the nuclear burning in the star should stop and for a good understanding of that, a proper theory of nuclear structure and nuclear fusion for the light nuclei was needed. Systematic studies in supernova explosions with inputs from nuclear and particle physics were started by different groups only in the 1960s and 1970s. Similarly, though Lev Landau in 1931 thought of a highly dense astrophysical object as a giant nucleus and the theory of neutron stars was worked out in the 1930s, the observation and compilation of neutron stars were done decades later. Particularly in the last few years, the observation of binary neutron star mergers along with the detection of gravitational waves and thermal plus non-thermal electromagnetic radiation has led to spectacular advances in the theory of neutron star matter and the process of heavy element nucleosynthesis.

This book describes the present knowledge one has about some of these key areas of nuclear astrophysics- the evolution of our understanding of the physics over the years and the recent developments in each of these areas.

In Chap. 2 the theory of core collapse supernovae is presented. In the initial stages of the collapse, electron capture by the iron-type nuclei reduces the pressure and makes the collapse faster. As the collapse proceeds, the density of matter

D. Bandyopadhyay, K. Kar, *Supernovae, Neutron Star Physics and Nucleosynthesis*, Astronomy and Astrophysics Library, https://doi.org/10.1007/978-3-030-95171-9_1

1

increases and at densities beyond 10^{12} g/cm^3, neutrinos and antineutrinos get trapped and cannot move out of the star. Then as the central density crosses the nuclear matter density, the matter becomes much stiffer centrally and the pressure there is dominated by the nuclear matter and not by the electron gas. The collapsing matter bounces back, the matter deep inside starts moving out and a shock wave forms at a radius where the infall velocity equals the outward velocity of the central region. This shock wave gets energized by pressure waves coming from inside and eventually travels outward. The chapter discusses why the moving shock cannot reach the edge of the core as seen in simulations done in one dimension, i.e. the radial direction. The nuclei ahead of the shock break into neutrons and protons and cause the shock to lose energy fast. But detailed physics inputs, taking care of the heating of the outer matter by neutrinos along with instabilities developed, are seen to make the shock move outward again and lead to an explosion on a longer timescale. This is seen in hydrodynamic simulations done in two and three dimensions, carried out in the last two decades and these results are briefly summarized in the chapter. The chapter also describes the details of the production and emission of a huge flux of neutrinos and antineutrinos of all flavors and their possible detections in the terrestrial detectors in the future. The first detection of the supernova neutrinos that took place in February 1987 coinciding with the observation of SN1987A is also presented.

Chapter 3 describes our knowledge of neutron stars and the physics involved in describing them. First pulsar which is a rotating neutron star with a strong surface magnetic field was discovered by Jocelyn Bell and Antony Hewish in 1967. By now there are about 3000 neutron stars known in our galaxy. Neutron stars are observed across the electromagnetic spectrum as well as in gravitational waves. With the new generation x-ray telescope NICER already in operation and the radio telescope SKA coming up and expected to start early science observations using a partial array around 2025, the observational study of neutron stars is entering into a new era. Observations of neutron stars lead to an estimate of masses (M), radii (R), the moment of inertia (I), surface temperatures, and magnetic fields of neutron stars. These observables particularly M, R, I are direct probes of compositions and equation of state (EoS) of matter in neutron stars. In this context, it should be mentioned that neutron stars are excellent celestial laboratories of fundamental physics under strong gravitational fields and extreme astrophysical conditions which could not be otherwise investigated in terrestrial laboratories. The compositions of neutron stars from crust to core are presented in Chap. 3 along with the different models that describe the EoS of the neutron star matter. One of the main focuses in this chapter is the study of dense nuclear matter, its EoS, and compositions from low to very high baryon densities relevant to supernovae and neutron stars using different effective models for strongly interacting matter. As the EoS is described in the parametric space of baryon density, temperature, and isospin, zero as well as finite temperature, effective theories are discussed in this chapter. Novel phases of the dense matter composed of hyperons, Bose–Einstein condensates of antikaons or quarks, phase transitions from one form of the matter to another form and their implications on the structures of (non-)rotating neutron stars, moment of inertia and quadrupole moment are described here. Further, the behavior of the dense matter in strong magnetic fields is highlighted in this chapter.

A separate chapter, chapter 4, deals with binary neutron star mergers and its connection to neutron star EoS using the zero and finite temperature effective theories. The 1967 discovery of the first pulsar paved the way for the discovery of the first binary pulsar PSR 1913+16 in 1974 by R. A. Hulse and J. H. Taylor. The Hulse–Taylor pulsar was later credited with the first indirect detection of gravitational waves (GWs). Einstein's theory of general relativity got another confirmation 100 years later when GWs from two colliding neutron stars were detected on 17th August, 2017. It was also the golden jubilee year of the first pulsar discovery. This was the first direct detection of GWs in a binary neutron star (BNS) merger event denoted as GW170817. This opened up another window in the study of neutron stars along with electromagnetic observations. It was possible to extract finite size effect, i.e. tidal deformation of one neutron star due to the other in the binary from the gravitational wave signal of GW170817. It is well known that the tidal deformability is sensitive to the cold EoS of neutron star matter particularly to the radius of a neutron star. This provides another opportunity to probe the EoS. The BNS merger event GW170817 was not only detected in gravitational waves, but was also observed across the electromagnetic spectrum. This gave birth to the multimessenger astrophysics along with GWs. The amount of material expelled in GW170817 did not support a prompt collapse of the remnant into a black hole. It was argued that the remnant in GW170817 survived for ~1 s and collapsed to a black hole thereafter. An upper bound on the maximum mass of neutron stars could be estimated in this scenario. This puts another constraint on the EoS along with the lower bound on the maximum mass of neutron stars as obtained from the observations of the galactic neutron stars. No gravitational wave signal at a few Kilo Hertz was detected from the merger remnant, a hot and neutrino-trapped compact object, due to the poor sensitivity of the present generation LIGO detectors in that frequency range. But the third generation gravitational detectors such as the Einstein Telescope and Cosmic Explorer would provide important information about the hot EoS, in future. In this connection, we discuss the imprints of hyperons and the hardon-quark phase transition on the gravitational wave signals.

Finally, Chap. 5 discusses the topic of nucleosynthesis through the $s-$, $r-$, and p-processes, the role of r-process in BNS mergers and the EoS of neutron star matter. It is predicted that r-process could be responsible for the synthesis of some of the heavy elements in supernova explosions and neutron star crusts under extreme physical conditions. The electromagnetic (EM) observations of ejected material in the BNS merger GW170817 found blue and red emissions. These features of the EM counterpart of GW170817 could be explained by the Kilonova (KN) model. The blue KN was the result of r-process nucleosynthesis in the ejected material with a high electron fraction >0.3 whereas the red KN was made of lanthanide nuclei synthesized in the neutron-rich ejecta via r-process with low electron fraction <0.2. The amount of material ejected in a BNS merger sensitively depends on the EoS of neutron star matter.

The extensive work done in these areas of nuclear astrophysics is helping the growth of both nuclear physics and astrophysics. The dense matter at moderate temperatures like a few tens of MeV and densities like a few times the saturation density will be explored in Facility for Antiproton and Ion Research, FAIR experiments at GSI. On the other hand more sensitive detectors for observing the gravitational waves and also the electromagnetic radiations from BNS or neutron star-black hole mergers are also envisaged in this decade.

Chapter 2
Theory of Supernova Explosions

Summary The physics of the spectacular core collapse supernova explosions starting with the collapse of massive stars with masses larger than 8 $M_\odot$ and ending with the birth of neutron stars or black holes is one of the most exciting areas of research for more than fifty years. The effect of neutral currents on the neutrino-matter interaction, the role of neutrinos and antineutrinos of all flavors in the energy loss, the equation of state of matter at high densities have all been explored in detail to throw light on the explosion mechanism. It became clear that once the central density becomes higher than the nuclear matter density, matter bounces back giving rise to a shock wave and the shock wave starts moving outward. However, almost all models, except some for the smaller masses of 8–10 $M_\odot$ fail to result in explosions when considered in one dimension, i.e. under the spherical symmetry. The cause is the depletion of energy through neutrino emission and nuclear dissociation, with the shock wave becoming an accretion shock. The importance of developing the hydrodynamics in numerical models in the other two angular dimensions is crucial. The effect of the heating of the outlying matter by a fraction of the huge flux of radiating neutrinos and the convection circles in the θ direction are found to contribute to the shock revival. A number of other effects and instabilities are also important and today some of the two- and three-dimensional hydrodynamic codes are able to produce the core collapse explosions with energies in the observed range around 10^{51} ergs. But still, these codes differ in some of the physics inputs and their effects in the final stage of the shock propagation. Thus a comprehensive understanding of the late stages of the core collapse explosion still awaits us.

2.1   Overview: Historical

Understanding the mechanisms of supernova explosions has been one of the most challenging problems of astronomy and astrophysics. After observing the sky for about two thousand years and with much more detailed investigations after the invention of the telescope, finally in the last century one connected the observation of supernovae to the death of large stars. In more recent years seeing a number of them each year, astronomers realized that there are two types of supernova type I and type II. Observationally type II shows strong lines of hydrogen while type I shows none. The light curves, meaning the optical luminosity as a function of time, also are different for the two types. The type II supernovae all start with the collapse of the inner part and the supernovae with this feature are called the core collapse supernovae (CCSN). Type I supernovae are subdivided into type Ia, Ib, and Ic. Type Ib and Ic also start with the collapse of the core. Today after large and sustained efforts of the last 50–60 years, one is able to explain the events that lead to the explosion starting with the collapse. Some uncertainties still remain in the understanding of the very late stages of the process. All these will be described in this chapter. Except for a short introduction to type Ia in the next section, we shall confine our attention to the core collapse supernovae only. But first, let us give a historical overview of observing the supernovae [1].

In the vast period of the last 2000 years or so, there are records of observation of a number of prominent supernova explosions. The earliest one goes back to 185 AD which still has a remnant giving x-ray images today. There are claims of observation of supernovae in the next few centuries but they are not confirmed as these have not yet been associated with any remnant. The supernova of 1006 AD seen in the constellation of Lupus, and watched from China and a few other countries, is still considered to be the brightest supernova. After that the event of 1054, recorded by the Chinese, Japanese, and the Arabs, had its remnant which is known as Crab Nebula. It was seen in the constellation of Taurus in the daylight for 23 days and in the sky at night for a long period of 653 days! The 1054 supernova left a pulsar, i.e. a rotating neutron star at the center in contrast to the events of 185 and 1006 which had no neutron star remnant.

In the year 1572, the astronomer Tycho Brahe observed in the constellation Cassiopeia in our Milky Way galaxy, a supernova. It was brighter than Venus and visible for several months. Tycho also saw that its position did not change relative to the fixed stars. The remnant of this supernova is still observed today in x-rays but there is no pulsar in it. After that in October 1604, three years after the death of Tycho Brahe, another bright starlike object appeared in the sky and this supernova could be seen for a full year. This one was observed by the astronomer Kepler, who was an assistant of Tycho Brahe. It was not as bright as the one observed by Tyco but other astronomers also observed it over a period of time and one could construct the light curve for this to some extent. One more supernova explosion took place in our galaxy, Cassiopeia A (Cas A) observed in 1680 by John Flamsteed, Astronomer Royal of England. But this event was not mentioned by other astronomers of that time and there is some uncertainty about some aspects of it. After that many supernovae have been observed in other galaxies but not yet in our own!

In the last fifty years as the facilities for the observation of supernovae, not only optical but over the whole range of electromagnetic radiation from radio to x-ray and γ-ray, improved a number of supernovae are observed every year. They are numbered by the year of observation followed by an alphabet. Of particular interest is the supernova SN1987A, seen on February 23, 1987. This was the first time a supernova was identified with the disappearance of a progenitor star Sanduleak -69 202 in the Large Magellanic Cloud (LMC) (except may be the strange case of SN 1961V [2]). The progenitor was a blue supergiant with a main sequence mass of 16 – 22 $M_\odot$ and a radius of $(3 \pm 1) \times 10^{12}$ cm, while a red supergiant would have a radius of about ten times more. The distance of LMC is 50 ± 5 kiloparsecs (kpc) and is quite close to our Milky Way. This supernova could be seen by the naked eye after the one of 1604 and was visible from the southern hemisphere. The SN was also unique as neutrinos emitted during the explosion were detected by two large water detectors, the Kamiokande II (K II) in Japan and the Irvine-Michigan-Brookhaven (IMB) in the state of Ohio, USA. The number of SN neutrinos detected were 12 and 8 spread over a time of about 12 and 6 seconds at K II and IMB, respectively. The observation of the neutrinos was considered a big success for the theoretical models for the physics of type II supernova in existence then, as they had predicted an associated flux of neutrinos roughly in agreement with the observed numbers. The optical display was approximately two to three hours after the arrival of the neutrinos to earth.

Presently the sensitivity and size of detectors of both the electromagnetic radiation and the neutrinos have increased many times compared to the time of SN 1987A. As a result supernova events from very large distances are observed now through optical and other telescopes each year, but no simultaneous emission of neutrinos has been detected yet.

2.2 Supernova Type Ia

Supernova of type Ia is due to thermonuclear disruption of white dwarfs. Observationally they do not have hydrogen lines but present a singly ionized silicon line near peak light. The type Ib or Ic have weak or no Si absorption feature. A white dwarf, mainly with carbon and oxygen and accreting mass from a companion star reaches a critical mass, and carbon, or possibly helium, gets ignited under degenerate conditions. The end product in Type Ia is Fe, which is produced after ^{56}Ni decays to ^{56}Co which in turn decays to ^{56}Fe. The half-life of the decay of ^{56}Co is 77 days. For quite a few observed cases one sees a half-life of 56 days and that is understood to be caused by the escape of the γ rays emitted by ^{56}Co, due to the increasing transparency of the material [1]. The type Ia supernovae are used as standard candles for determining the absolute magnitude of galaxies and thus their distance [1]. This is because they arise from a consistent type of progenitors which after mass acquisition explode when they reach a typical mass. The peak luminosity of the light curve is typically seen as having a visual absolute magnitude of about −19.3 (about 5 billion times brighter than the sun). This similarity of behavior in them leads to their usefulness in distance measurements and has been used in some very important projects like the discovery of the universe accelerating outwards.

2.3 Gravitational Collapse and Pre-supernova Conditions

The large number of supernova events that are observed depends roughly on three important overall parameters -the mass, the metallicity, and the rate of rotation of the stars [3]. Let us consider the simpler situation of no rotation and no mass loss. According to Woosley and Heger [3], there are several mass ranges for which the "outcomes" are seen to be qualitatively different. The lower range of 8 to $30 \, M_\odot$ for the main sequence stars has roughly the pre-supernova helium core masses up to $12 \, M_\odot$ and they end up with the iron-type nuclei at the center once the fusion reactions are over.

Let us look more closely at the different stages of nuclear burning. The stars after being formed from interstellar dust go through the fusion reactions and for given temperature and composition, the reactions with the lowest Coulomb barrier proceed most rapidly. These exothermic reactions also provide the internal pressure that balances the gravitational pressure. Figure 2.1 gives the different burning phases of a spherically symmetric star of mass $25 M_\odot$. The hydrogen burning in stars like the sun lasts for about 7–10 billion years. But the timescale of the hydrogen burning in the larger stars of mass $25 \, M_\odot$ turns out to be about 7 million years, as shown in Fig. 2.1, with the typical central density as $5 \, \text{g/cm}^3$ and a central temperature of $6 \times 10^7 \text{K}$. At the end of it, when the hydrogen in the central region is all converted to helium, the star contracts under gravity. As a result the temperature rises and that eventually triggers the burning of the helium. That again continues for about 0.5 million years for the $25 M_\odot$ star as shown in Fig. 2.1 with the density and temperature being $700 \, \text{g/cm}^3$ and 2×10^8 K, respectively. But the hydrogen burning still continues in an annular shell outside the central He region. This is followed by carbon burning where C/O fuse to give O/Ne/Mg in a shorter timescale of about 600 years as shown in Fig. 2.1. These finally end up with the silicon getting converted to Fe-type nuclei mostly by repeated capture of alpha particles in a timescale of a day to a month. The products like the nuclei ^{56}Fe and ^{56}Ni have the largest value for binding energy per nucleon and thus fusion cannot result in the release of energy anymore. The star at this stage is often thought of as having an onion skin structure, as shown in Fig. 2.2. The central region of Fe-type elements have an outer annular region where fusion reactions go on with Si and S, with a region where O and Ne burning taking place further outside, with C burning in the shell outside of it, with another He burning shell and finally H burning shell up to the surface of the star. The Si burning shell becomes thinner with time, adding mass to the stellar core of the iron-type nuclei.

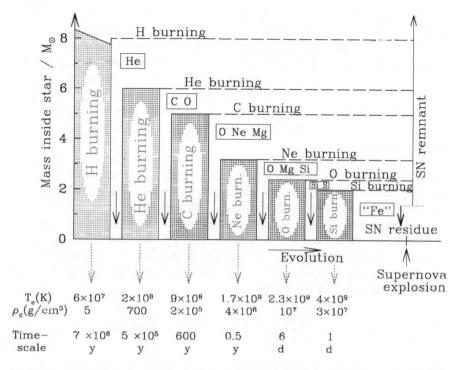

Fig. 2.1 Burning stages of stars with their densities, temperatures, and timescales of a typical spherically symmetric star of mass 25 $M_\odot$. This is taken from [4] and reproduced by permission of the IoP Publishing

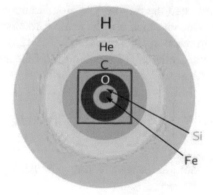

Fig. 2.2 Onion skin structure of a typical 25$M_\odot$ star at the end of silicon burning in the core. This is taken from Ref.[5] which is an open access article distributed under the terms of the Creative Commons Attribution License

Before going into the details for the stellar mass range of 8 to 30 $M_\odot$, let us briefly mention the outcomes of other heavier stars. According to Woosley and Heger [3], for the stars in the mass range of 30 to 80 $M_\odot$ with helium core mass of 10 to 35 $M_\odot$, one generally expects the formation of black holes whereas for stellar masses below 30 $M_\odot$ one mostly gets neutron stars as remnants of the explosion [3]. There are similar predictions for stars having masses more than 80 $M_\odot$. Also the role that the pair instability plays at the advanced stages of the evolution of some of these very massive stars leading to the supernovae known as "pair instability supernovae," has been studied [3].

Near the final stages of the Si burning phase, the nuclear statistical equilibrium (NSE) is reached, which means that the rates of all strong and electromagnetic reactions are equal to the rates of their inverse processes. This steady state implies a considerable simplification in the calculation of the abundances of the nuclei present in the core, as we see below. In this situation one can use the statistical principles that are used for the Saha ionization formula [6, 7] for the ionization of atoms. For the breakup reaction of a nucleus (N, Z) with neutron number "N" and proton number "Z," given by $(N, Z) \rightarrow (N - 1, Z) + n$, the ratio of the number densities $\frac{n(N-1,Z)}{n(N,Z)}$ is related to the temperature "T" by

$$\frac{n_n n(N - 1, Z)}{n(N, Z)} = \left[\frac{2G(N - 1, Z)}{G(N, Z)}\right] \left[\frac{(2\pi \mu kT)^{3/2}}{h^3}\right] exp(-Q_n/kT) , \quad (2.1)$$

where $G(N, Z)$ stands for the partition function for the nucleus (N, Z) and Q_n for the binding energy of a neutron in the nucleus (N, Z), a positive number for stable nuclei. Also "k" is the Boltzmann constant and "h" is the Planck constant. The quantity "μ" is the reduced mass of the system of the nucleus $(N - 1, Z)$ and the neutron. We denote by $M(N, Z)$ the mass of the nucleus (N, Z) and by M_n, M_p the masses of the neutron and the proton. The atomic number of the nucleus is A and equal to $(N + Z)$. One can then use the same formalism of the nuclear Saha equation for the breakup reaction $(N - 1, Z) \rightarrow (N - 2, Z) + n$, and so on, and after a series of equations, get the final form in terms of the neutron number density n_n and proton number density n_p [8, 9]

$$n(N, Z) = G(N, Z) A^{3/2} \left[\frac{(n_n)^N (n_p)^Z}{2^A}\right] \theta^{(1-A)} exp(Q(N, Z)/kT) , \quad (2.2)$$

where $\theta = \frac{(2\pi M_u kT)^{3/2}}{h^3}$ with M_u the atomic mass unit. Also

$$Q(N, Z) = [ZM_p + NM_n - M(N, Z)]c^2 , \quad (2.3)$$

and equal to the binding energy of the nucleus (N,Z). The Eq. (2.2) for the nuclear statistical equilibrium still needs two more constraints as the number densities, n_n and n_p are not known. The first of these supplementary equations is essentially the conservation of mass. For a collection of nuclei with masses M_i and number densities n_i (abundances X_i), having N_i neutrons and Z_i protons, it is given by

$$\Sigma_i X_i = 1 \quad or, \quad \Sigma_i n_l M_i = \rho N_{Av} , \tag{2.4}$$

where N_{Av} is the Avogadro number. The sum in Eq. (2.4) over all nuclei includes free neutrons, protons as well as alphas. For the second, we observe that though the rates of weak interaction reactions, like the electron capture on nuclei and its inverse process the neutrino capture are not in equilibrium, their time scales are much larger and as a result that does not affect the abundances much. Then the total number densities of free and bound protons and neutrons preserve the neutron excess, η and hence

$$\Sigma_i X_i (N_i - Z_i)/M_i = \Sigma_i n_i (N_i - Z_i)/\rho N_{Av} = \eta . \tag{2.5}$$

We shall use instead of η, the electron fraction Y_e which is defined as the total number of electrons divided by the total number of baryons. It is then by charge conservation, equal to the ratio of the total number of protons and the total number of baryons and related to η by $Y_e = (1 - \eta)/2$. This Y_e is a slow function of time and can be taken a constant for the abundance calculations. Thus all the abundances are characterized by the three quantities density, temperature, and Y_e.

The stellar cores during the evolution have pressure inward due to gravitation and outward pressure due to the relativistic degenerate electron gas and the radiation pressure due to the nuclear reactions. Once the nuclear reactions stop after conversion of Si to Fe/Ni in the stellar core, the electron pressure alone cannot counter the inward gravitational pressure for core masses beyond a critical value. This was observed by S. Chandrasekhar as early as 1931 [10, 11]. So stars that develop core masses larger than this critical mass, called the Chandrasekhar limit, start collapsing. As the pressure of the relativistic electrons is related to the electron fraction Y_e, the Chandrasekhar limit mass [8, 10, 11] is found to have the following form [1]

$$M_{Ch} = 5.8 Y_e^2 M_\odot . \tag{2.6}$$

So before the beginning of the collapse, known as the pre-supernova stage, the matter consists of iron-type nuclei, along with a small fraction of free neutrons, a much smaller fraction of free protons in equilibrium with the decaying free neutrons and of course, the gas of relativistic electrons. The nuclear statistical equilibrium also determines the abundances of the other elements and their isotopes. The electron captures (EC) on the nuclei and the free protons in the pre-supernova stage are given as

$$(N, Z) + e^- \quad \rightarrow \quad (N + 1, Z - 1) + \nu_e , \tag{2.7}$$

$$p + e^- \quad \rightarrow \quad n + \nu_e , \tag{2.8}$$

They destroy the electrons and reduce the electron fraction Y_e. That correspondingly reduces the mass of the core that can withstand the inward gravitational pressure. On the other hand, if some isotopes undergo beta decay that increases the electron fraction and helps in countering the gravitational pressure and the beginning of the collapse.

So one realizes that the electron capture and the beta decay reaction rates for the nuclei with atomic numbers up to 56 and somewhat higher need to be known. This motivated nuclear physicists to perform structure calculations using shell model and other phenomenological models [12, 13] for nuclei heavier than ^{40}Ca, known as the fp-shell nuclei. As the nuclei are inside a star with a non-zero temperature of roughly 1 MeV, the rates and half-lives get modified from their known laboratory values. This is an interesting example where the microscopic physics of nuclear states and their transition strengths by weak interaction processes decides the beginning of stellar collapse and the size of the core.

The typical values for the density, the temperature, and the proton-to-baryon ratio, Y_e at the start of the collapse are needed to be known for further evolution. There is general agreement in these values among different groups. Janka et al. [5] gives a central temperature of 10^{10} K or about 1 MeV, a central density of several 10^9 g/cm^3 and Y_e of 0.45 with a typical diameter of 3000 km.

The other physical quantity that plays an important role throughout the collapse is the entropy. The entropy per nucleus for translational motion is given by Bethe et al. [12]

$$\frac{S_{nucl}}{\bar{N}k} = \frac{5}{2} + ln \, [(\frac{MkT}{2\pi\hbar^2})^{3/2} \, (\frac{V}{\bar{N}})] \, , \tag{2.9}$$

where $\bar{N}$ and M are the number and average mass of the nuclei in a volume V and T is the temperature. One can similarly calculate the entropy of the alpha particles and neutrons present, produced through thermal dissociation with their abundances given by the nuclear statistical equilibrium. The electron Fermi gas has an entropy per electron, equal to

$$S_e = \frac{\pi^2 kT}{\epsilon_F} \, , \tag{2.10}$$

where ϵ_F is the Fermi energy of the electron gas. Numerically, assuming the nuclei typically as ^{56}Fe one finds that for a typical temperature of 8×10^9 K the nuclei have entropy per nucleus of about 16.7. At that temperature, the nuclei have a probability of getting their excited states occupied. Numerically that gives an additional contribution of about 4.8 per nucleus [12]. The alphas and neutrons contribute an entropy of about 3.6 to that and the electron gas with a Fermi energy of 6 MeV has entropy of 1.1 per electron at that temperature [12]. Adding all these contributions the total entropy, (S/k) per nucleon turns out to be a bit less than 1. So the important point to note is that at the start of the collapse, the matter in the form of nuclei has a small value of entropy, about 1 per nucleon. As the collapse proceeds, the entropy changes. We shall estimate that change and its consequences in the next section.

2.4 Production of Neutrinos and Their Emission

The collapse is initiated by photodissociation of some nuclei to alpha particles as well as by electron captures on the nuclei. As the electron captures reduce the electron pressure, matter falls in and the density increases with time. The electron type neutrinos that get liberated by the captures hardly interact with the matter, at these densities of 10^9 g/cm^3 to a few times 10^{11} g/cm^3, and escape from the star. So do the electron type antineutrinos that are created by the inverse process of beta decay. Early work on the collapse with an analytical collapse rate was done assuming spherical symmetry. But as we shall discuss later in this chapter, this treatment in one dimension, i.e. involving the variable, the radial distance, is found inadequate and does not give the right results. One needs to consider the second variable, the angle "θ" or even both the angles "θ" and "ϕ" and then proceed dividing the area or the volume into specific zones and then writing the equations of motion in each zone, solve them numerically with continuity on their interfaces. These are the so-called hydrodynamic codes for computation. But for the initial stage, we shall describe the physics assuming spherical symmetry. Then using Newtonian gravity, one can write for a mass element at a radial distance "R" from the center of the core, the equation of motion [1]

$$\frac{d^2R}{dt^2} = -\frac{GM_{inside}}{R^2} - (\frac{1}{\rho})\frac{\partial P}{\partial R} , \qquad (2.11)$$

where M_{inside} is the total mass inside the sphere of radius R, "ρ" is the density, G the gravitational constant and $P = P(\rho, T, Y_e)$ is the pressure, a function of the density, the temperature and the electron fraction. This can be used for lower densities at the early stages of collapse but when densities become very high, one needs to use the general relativistic form for the equation.

Goldreich and Weber [14] and later Yahil and Lattimer in [15] and [16], observed that the inner part of the core collapses in a self-similar manner maintaining a "structural integrity," with the falling velocity proportional to the distance from the center. This is called a homologous collapse where all hydrodynamic variables like the density, the velocity, the mass inside the core can be written in terms of a "similarity variable" [17]. Arnett [17] compared the infall velocity and the sound speed for a typical collapse and concluded that about 0.5–0.7 M$_\odot$ of the core formed this homologous core at the center.

For an understanding of the physical processes that occur while the density increases by several orders of magnitude, one needs a collapse rate as a function of the increasing density and this is written as [1]

$$\frac{d(ln\rho)}{dt} = 200 \, \rho_{11}^{m \, 1/2} \, , \tag{2.12}$$

where ρ_{11}^m is the density at half the mass of the homologous core, in units of 10^{11} g/cm^3.

Now let us turn to a detailed discussion of the weak interaction processes, the beta decay, and the electron capture rates of nuclei inside stars. The β^- decay half-life of a nucleus in the laboratory $t_{1/2}$ is given by $\frac{ln2}{\lambda_{\beta^-}}$. The decay rate λ_{β^-} for the ground state (gs) of the mother nucleus can be written as

$$\lambda_{\beta^-}(gs) = \frac{ln\,2}{t_{1/2}(gs)} = (\frac{ln\,2}{K})\Sigma_j \, [B_j^F + (\frac{g_A}{g_V})^2 B_j^{GT}] f(Z, Q_0) \, . \tag{2.13}$$

The constant $K = \frac{ln\,2}{g_V^2} \, [\frac{2\pi^3\hbar^7}{m_e^5 c^4}] = 6146 \pm 6$ s with m_e the electron mass, $\hbar = \frac{h}{2\pi}$ with "h" the Planck's constant, and c the velocity of light. g_A and g_V are the axial vector and vector coupling constants of weak interaction. In Eq. (2.13), the sum over "j" includes all the states of the daughter nucleus that are reached by the Q-value of the decay, Q_0. For the β^- decay, the quantities B_j^F and B_j^{GT} are the strength of the Fermi operator ($\Sigma_i t_+(i)$) and of the Gamow-Teller (GT) operator ($\Sigma_i \sigma(i)t_+(i)$) between the ground state of the mother and the state $|j>$ of the daughter nucleus. Here $\mathbf{t} = (1/2)\tau$ with τ standing for the three Pauli matrices in the isospin space. When the decaying nucleus is in a stellar core at a temperature "T," the excited states of the mother nucleus, with energies E_i, that get populated thermally also make contributions and the expression for the rate gets modified to

$$\lambda_{\beta^-} = \left(\frac{ln2}{K}\right)\left(\frac{1}{G}\right) \Sigma_i(2J_i + 1)exp(-E_i/kT)$$

$$\times \Sigma_j \left[B_{ij}^F + (\frac{g_A}{g_V})^2 B_{ij}^{GT}\right] f(Z, Q_i), \tag{2.14}$$

where "i" sums over the ground and the excited states of the mother nucleus. Assuming isospin conservation, the Fermi operator for the β^- decay connects each mother state with isospin "t" to the Isobaric Analog State (IAS) in the daughter. The isopin selection rule for Fermi transition is $\Delta t = 0$ and the total Fermi strength is $B_F = [t(t + 1) - m_t(m_t + 1)]$ where m_t is the eigenvalue of the operator t_z, the isospin z-component of the mother. The GT operator for β^- decay connects one state of the mother nucleus to a number of states of the daughter with final isospin $t' = t - 1, t$ or $t + 1$ and so the GT strength has a distribution over the energies of the discrete final states. When many states contribute, this is often given as a continuous

distribution called a giant resonance. The GT strength when summed over all final
states satisfy a simple sum rule given by $S_{\beta-}$ -$S_{\beta+}$ = 3 (N-Z) where $S_{\beta-}$ is the sum
of the B^{GT} for the β^- decay of the mother nucleus (N, Z) and $S_{\beta+}$ for its β^+ decay
(operator for β^+ decay GT transition is $\Sigma_i \sigma(i) t_-(i)$). The phase factor $f(Z, Q_0)$
for ground state β^- decay is given by

$$f(Z, Q_0) = \int_1^{\epsilon_0} F(Z, \epsilon)(\epsilon_0 - \epsilon)^2 \epsilon(\epsilon^2 - 1)^{1/2} d\epsilon , \qquad (2.15)$$

with $\epsilon = \frac{E_e}{(m_e c^2)}$ where E_e is the electron energy variable, $\epsilon_0 = \frac{Q_0}{(m_e c^2)}$ and $F(Z, \epsilon)$
is the Coulomb distortion factor [18] that takes into account the distortion of the
emitted electron wave function by the nuclear charge. For a transition from the i-th
state of the mother, with the Q-value given by $Q_i = Q_0 + E_i$, the upper limit of the
integration gets changed to ϵ_i where $\epsilon_i = Q_i/(m_e c^2)$. For large values of Q_i and
ignoring $F(Z, \epsilon)$, the dominant term goes as the fifth power of Q_i.

For the β^+ decay and the electron capture, the transition is from initial isospin "t"
to final isospin "$t + 1$." The Fermi has no contribution and the total GT contribution
is much smaller compared to the β^- decay case. Thus the $S_{\beta-}$ is marginally higher
than the value 3 (N-Z). The phase space factor for the electron capture is very
different from the beta decay case, as there are two particles in the final state instead
of three [18].

Normally for stable nuclei to capture electrons, the electrons gas must have
a Fermi energy greater than -Q MeV, i.e. greater than the difference of binding
energies of the daughter and the mother nuclei. But as with the collapse, the density
goes up from the value of 10^9 g/cm^3, to 10^{12} g/cm^3. the electron chemical potential
increases from a few MeV to 30–40 MeV. This makes faster electron captures
possible reducing Y_e from its initial value of 0.45. In the pre-supernova and early
collapse, β^- decays oppose electron captures and increase the value of Y_e. But soon
this stops contributing because once the electron Fermi energies become large, the
electron created in the decay has to go to the top of the Fermi sea. As the energy
released in the β^- decays of the relevant nuclei are a few MeV to 10–12 MeV, the
process gets blocked.

For both the electron capture rates and the beta decay rates, one needs to know
the Gamow-Teller strength distribution as a function of the excitation energy of
the daughter nucleus. As mentioned earlier this motivated one to perform nuclear
structure calculations like the shell model. The first shell model estimates were
provided by Fuller, Fowler, and Newman [19, 20] and [21]. Later different groups
improved the calculations. At one time the mid-fp shell, i.e. for nuclei with an
atomic number around 60, shell model calculations with all the valence nucleons
were prohibited by demands of very large space matrix diagonalization. But that
was overcome by the development of the code ANTOINE by E. Caurier and beta
decay and electron capture rates for nuclei with atomic number up to $A = 65$ were
calculated which agreed with experimentally observed strengths for the low-lying
states [22, 23]. Another development that took place was the formulation of the

Shell Model Monte Carlo (SMMC) approach [24], which was able to calculate the thermally averaged properties of nuclei, in extremely large spaces going beyond the full fp shell. This was then combined with the β^+ decay GT strength distributions calculated in the Random Phase Approximation (RPA) framework to produce the electron capture rates for a large number of fp-shell nuclei for appropriate collapse conditions.

Experimentally laboratory beta decays give the β^- GT strengths for the excitation energy in the daughter only up to the Q-value. But over the years the charge exchange reactions like (p, n) reactions for β^- and (n, p) reactions for β^+/EC, were carried out in different nuclear reactions facilities to find the GT strength distributions over the entire range of excitation energy. Later the other set of charge exchange reactions $(^3He, t)$ and $(d, ^2He)$ also gave the required strengths. But one has seen consistently from these experimental results that the theoretically calculated GT strength sums overestimate the observed strengths integrated over the whole range of energy of the daughter nucleus, by an approximately constant factor called the "quenching factor." So for a correct evaluation of the reduction of the value of the electron fraction Y_e, one needs to use the quenched GT strengths for calculating the rates.

As the collapse proceeds, the next important physics effect takes place at a density of about 10^{12} g/cm^3. At this density, the neutrinos and antineutrinos that were earlier escaping the star, start staying inside as their mean free path becomes of the order of the core radius. let us first enumerate the weak interaction processes active during the collapse phase [25–27]

$$p + e^- \;\rightleftharpoons\; n + \nu_e , \tag{2.16}$$

$$n + e^+ \;\rightleftharpoons\; p + \bar{\nu}_e , \tag{2.17}$$

$$(N, Z) + e^- \;\rightleftharpoons\; (N + 1, Z - 1) + \nu_e , \tag{2.18}$$

$$(N, Z) + e^+ \;\rightleftharpoons\; (N - 1, Z + 1) + \bar{\nu}_e , \tag{2.19}$$

$$\nu_\alpha + N_{nucl} \;\rightleftharpoons\; \nu_\alpha + N_{nucl} , \tag{2.20}$$

$$N_{nucl} + N_{nucl} \;\rightleftharpoons\; N_{nucl} + N_{nucl} + \nu_\alpha + \bar{\nu}_\alpha , \tag{2.21}$$

$$\nu_\alpha + e^\pm \;\rightleftharpoons\; \nu_\alpha + e^\pm , \tag{2.22}$$

$$\nu_\alpha + (N, Z) \;\rightleftharpoons\; \nu_\alpha + (N, Z) , \tag{2.23}$$

$$\nu_\alpha + (N, Z) \;\rightleftharpoons\; \nu_\alpha + (N, Z)^* , \tag{2.24}$$

$$e^+ + e^- \;\rightleftharpoons\; \nu_\alpha + \bar{\nu}_\alpha , \tag{2.25}$$

$$(N, Z)^* \;\rightleftharpoons\; (N, Z) + \nu_\alpha + \bar{\nu}_\alpha . \tag{2.26}$$

In these equations ν_α stands for a neutrino or an antineutrino of any flavor, $(N, Z)^*$ stands for the nucleus excited and e^+ for the positrons. Also in Eq. (2.20) and Eq. (2.21) "N_{nucl}" stands for a nucleon as distinct from the nucleus (N,Z).

The elastic scattering of the neutrino on nuclei given by Eq. (2.23) is mainly responsible for the neutrino trapping. The phenomenon of neutrino trapping was pointed out by Friedman [29], Mazurek [30] and Sato [31]. The neutral weak current between neutrinos and neutrons and protons (to a lesser extent) gives rise to the coherent elastic scattering of the neutrinos by nuclei. The theory was worked out by Tubbs and Schramm [32] and simplified versions of that followed later [12, 33].

The mean free path of neutrinos for this neutral current is given by Bethe et al. [12]

$$\lambda_\nu = 1.0 \times 10^6 \, (\frac{\rho}{10^{12} g/cm^3})^{-1} [(N^2/6A)X_h + X_n]^{-1} (\frac{E_\nu}{10 \, MeV})^{-2} \, cm \,, \quad (2.27)$$

where X_h and X_n are the mass fractions in the stellar material of the heavy nuclei and the neutrons, respectively, and E_ν is the neutrino energy in MeV. The value of $sin^2\theta_c$ with θ_c the Cabbibo angle, is taken as 0.25 in Eq. (2.27). If the more accurate value of 0.23 is used, one gets the factor $(N - 0.08Z)^2$ instead of N^2 in the RHS of Eq. (2.27) showing an explicit mild dependence on "Z." Below we give the arguments of Bethe [1] to show that at densities of 10^{12} g/cm^3, the neutrinos are indeed trapped.

We neglect the small fraction of free neutrons, i.e. take $X_n = 0$ and take the average value of "N" at density 10^{12} g/cm^3 as 50 (the average nucleus becomes heavier as collapse proceeds) and take Y_e as 0.4, i.e. N/A=0.6. Then Eq. (2.27) gives $\lambda_\nu = 2$ km$(10 \, MeV/E_\nu)^2 = 0.4$ km where average energy at this density is taken as 22 MeV [12, 34]. The diffusion length of the neutrinos goes as the square root of the time and given by $L = (c\lambda_\nu t/3)^{1/2}$ which is about 8 km (the time to reach the density 10^{12} g/cm^3 from onset of collapse is taken as 1.6 ms using the collapse rate, Eq. (2.12)). The typical radius of the core at this density is 30 km. Thus at densities of 10^{12} g/cm^3, the neutrinos do not diffuse out much.

In this connection, one defines the optical depth (τ) in terms of the radial coordinate as

$$\tau = \int dr/\lambda_\nu \,. \quad (2.28)$$

Then we can talk of a neutrino sphere from whose surface neutrinos can freely escape to infinity and Wilson defined its radius R_ν as

$$\tau(R_\nu) = 2/3 \,. \quad (2.29)$$

The value (2/3) is chosen rather than one, to take into account of the possibility of neutrinos not moving out radially [1]. This neutrino sphere is much further out compared to the trapping radius. Also one observes that the neutrino sphere for the electron type neutrinos has radius larger than the ones for ν_μ or ν_τ as they interact both through charged current and neutral current whereas the ν_μ or ν_τ have only neutral current interaction.

The thing to note here is that after neutrino trapping, though electron captures reduce Y_e, but the total lepton fraction $Y_l = Y_e + Y_\nu$ remains constant where Y_l and Y_ν are the total lepton number and the total neutrino number per baryon. This, of course, is at densities where thermal production of $\nu\bar{\nu}$ still does not occur.

The scattering of the neutrinos by the electrons results in the neutrinos losing energy, as the electrons are highly degenerate and hence can only gain energy. With lower energies, the neutrinos can escape more easily. The effect of this electron scattering on reducing the Y_l value before complete trapping and the change of entropy due to this, have both again been estimated by Bethe [1] and others.

One can also evaluate the change of entropy ΔS that takes place when the collapse proceeds and electron capture/beta decay changes the fraction of different species. One can use the thermodynamic relation

$$T dS = dQ - \Sigma_i \mu_i dN_i , \tag{2.30}$$

where μ_i stands for the chemical potential of the i-th species. For the example of an electron capture on a free proton with the emitted neutrino escaping, one sees that $dN_e = dN_p = -dN_n = -1$, and $-dQ$, the energy carried off by the neutrino is found by averaging the cross section ($\propto E_\nu^2$) over the neutrino momentum space and that comes out as $(5/6)\mu_e$. Thus with $\hat{\mu} = \mu_n - \mu_p$, the entropy change is given by

$$T\Delta S = \mu_e/6 - \hat{\mu} . \tag{2.31}$$

Equation (2.31) has the right hand side nearly always negative showing that the entropy getting somewhat reduced before neutrino trapping sets in. Of course after the neutrino trapping ΔS becomes positive. Though we do not calculate here the entropy change for each of the other processes involved in the post-trapping stage, they can similarly be calculated using the thermodynamic relations.

The collapse proceeds after the neutrinos get trapped with the inner part of the core moving inward with a velocity proportional to the radius whereas the outer part is falling in with the velocity proportional to $1/\sqrt(r)$ slightly higher than half of the free fall velocity. Very far out, the velocity is smaller because those parts are yet to receive the signal that the matter is falling in!

2.5   Shock Wave Formation and Its Eventual Stalling

During the infall, the pressure is essentially due to the leptons, initially it is the pressure due to the relativistic electron gas and after neutrino trapping, due to the electron and the neutrino gas. But when the density reaches the saturation nuclear matter density ρ_s, i.e. 2.7×10^{14} g/cm^3 or 0.16 fm^{-3}, the situation changes. The nuclei start touching each other and matter becomes much stiffer.

One of the most important quantities to study for the supernova collapse is the behavior of the pressure as a function of the density, i.e. the quantity $P(\rho)$. The adiabatic index Γ, defined as

$$\Gamma = [\partial(log P)/\partial(log\rho)]_S , \qquad (2.32)$$

at fixed entropy S, tells us how strongly the pressure increases with the density. Normally one talks of two definite stages for the Equation of State (EoS) i) below the nuclear matter density and ii) above the nuclear matter density.

At densities below the nuclear matter density, one considers the matter consisting of the nuclear and the electron component (after the neutrino trapping the lepton component). With nuclear and electromagnetic forces in action, statistical mechanics is used to derive the EoS. Because of attractive nuclear forces the nuclear component is condensed into nuclei and a low density gas of alpha particles and nucleons remains outside. Different groups like Lattimer et al. [35], Ravenhall, Pethik and Wilson [36], Cooperstein [37] derived the EoS in different forms, but all of them showed the following important feature: the nuclear contribution to $P(\rho)$ are seen to be small compared to that of degenerate electrons/leptons. It was also seen that Γ remained less than 4/3 up to densities close to the saturation density for the system, with the entropy per nucleon having values in the range of 1–2 and the collapse staying mostly homologous.

The situation at densities higher than the nuclear matter density is more difficult to handle theoretically. One of the important quantities here is the compression modulus of symmetric nuclear matter, defined as

$$K_{00} = 9(\frac{dP}{d\rho})_{\rho=\rho_0} . \qquad (2.33)$$

Baron et al. [38, 39] postulated a 'schematic' EoS for the nuclear matter at high density, given by Bethe [1]

$$P = \frac{K_0 \rho_0}{9\gamma}[(\rho/\rho_0)^\gamma - 1] , \qquad (2.34)$$

where K_0 is the compression modulus of the nuclear matter (not necessarily symmetric matter), ρ_0 is the normal density of the nuclear matter and γ a parameter. This EoS is often used for the computation of the supernova evolution. The modulus K_0 and the density ρ_0 are often parametrized as functions of the proton fraction $x(= Z/A)$ as [1]

$$\rho_0(x) = 0.16\left[1 - 0.75(1 - 2x)^2\right] fm^{-3}, \qquad (2.35)$$

and

$$K_0 = K_{00}\left[1 - 2(1 - 2x)^2\right].$$ (2.36)

Both of the above equations hold for small values of $(1 - 2x)$ only. Of course for the total pressure, one needs to add a thermal pressure P_{therm} to the expression for pressure in Eq. (2.34) with

$$P_{therm} = \frac{\pi^2 T^2}{6\epsilon_F}\frac{m^*}{m}\rho\, u^{-2/3},$$ (2.37)

where $u = \rho/\rho_0$, ϵ_F the Fermi energy of the electron gas (with a typical value like 34 MeV) and m^* and m the effective mass inside and the true mass of the nucleon, respectively. The value of the compression modulus was initially determined from the experimental observation of the breathing mode for doubly closed magic nuclei like ^{40}Ca, ^{90}Zr and ^{208}Pb and then extrapolated to the symmetric nuclear matter by Blaizot et al. [40] with the value

$$K_{00} = 210 \pm 30\, \text{MeV}.$$ (2.38)

However then Brown [41] argued that if the difference of nuclear matter and doubly magic nuclei is properly taken into account, the value of K_0 for ^{208}Pb should be 140±20 MeV, in agreement with the calculations of Pines [42].

But recently Stone et al. reanalyzed iso-scalar giant monopole resonance data based on the liquid-drop model description of vibrating nuclei and found the value of K_{00} in the range 250–315 MeV [43]. This is higher than the present-day accepted values of 248±8 [44] MeV and 240±20 [45] MeV.

So the question is what happens physically to the matter at the center of the collapsing core when the density goes beyond the nuclear matter density. Pressure starts building up and the infall velocity deep inside soon falls to zero. Figure 2.3 gives a schematic picture of the different stages of the supernova evolution [28]. The top left figure shows the collapse of the Fe core starting and the figure below that in the middle panel shows the "core bounce" at the center of the core with pressure waves travelling outward and as we discuss below, this gives rise to a shock wave. The diagram in Fig. 2.3 top right shows the phenomenon of neutrino trapping at densities $\sim 10^{12}$ g/cm^3.

We already mentioned that inside a homologous core, the velocity of the infalling matter is proportional to the distance from the center. On the other hand, the local velocity of sound decreases with the radial distance. So at some particular value of the radius, the speed of sound equals the infalling velocity and this point is known as the sonic point. A disturbance inside there has no influence beyond the sonic point and the waves there are stationary compared to the center of the core. Thus once the density goes beyond the "core bounce" density, pressure waves from inside bringing kinetic energy come and stop at the sonic point. This is because the

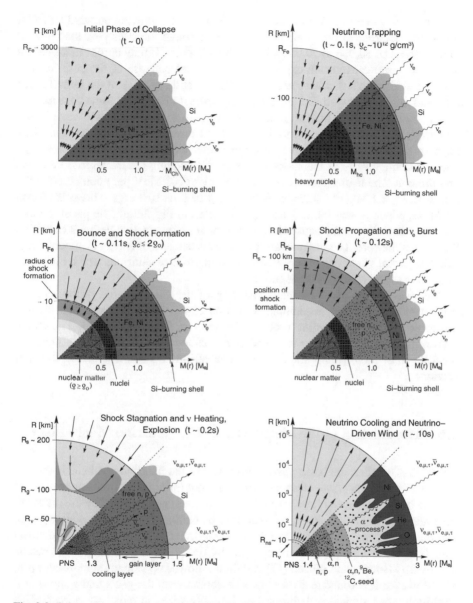

Fig. 2.3 Schematic representation of the different stages of a core collapse supernova explosion. Reprinted from [28], ©2007 with permission from Elsevier

speed of sound reduces going out and also because as they are moving up, matter flow inward is becoming faster with a larger radius. Within a millisecond or so, the waves collecting there bring enough energy to show a sharp rise and subsequently a discontinuity in the velocity. This discontinuity in the velocity is the shock wave.

The shock wave, once it gathers enough energy, starts moving outward fast with a velocity determined by its energy. It moves through the overlying matter with initial speeds of 30,000 to 50,000 km/sec. At this point, it may appear that one has got the supernova explosion explained, as an initial collapse of the core has got reversed leading to a shock wave created roughly at the surface of the homologous core and moving outward. But unfortunately, except may be for some core masses of smaller stars of $8-10\,M_\odot$, all other stars are seen to have the shock wave losing its energy and becoming an accretion shock, at a radius of 100 to 200 km. This happens because the shock moves through matter which still consists of iron-type nuclei and the energy of the shock is lost in dissociating the nuclei into neutrons and protons. The matter thus extracts an energy of 8.8 MeV per nucleon from the shock. Every $0.1\,M_\odot$ of matter needs an energy of 1.7×10^{51} ergs to break it up into nucleons, which is very large. This stage is shown in Fig. 2.3 middle panel left and right diagrams. Figure 2.3 bottom left shows the shock stagnation before it can reach the edge of the core. The fragmentation also leads to a loss of energy in another way. The free protons created, capture electrons, and produce neutrinos which come out of the star at that radius and thus cool it.

This was the status of the understanding of the core collapse process in the early 1980s. After that, explaining how one can still get an explosion, turned out to be a really challenging problem for the last forty years or so. The shock revival mechanism that came to the rescue is called the delayed-neutrino heating mechanism or more simply the neutrino mechanism. We describe that in the next section.

2.6 The Revival of the Shock Wave- the Neutrino Mechanism

Let us first mention the basic physics involved in the neutrino mechanism for the revival of the shock wave. Neutrinos from the hot inner core of the supernova get emitted in huge numbers. A very tiny fraction of them get absorbed by the matter in the region of the stalled shock, but that may be enough to revive the stalled shock and eject the outer part of the star leaving the inner mass to turn into a neutron star. This was first proposed in the analysis by Bethe and Wilson [46] of the results obtained by Bowers and Wilson [47] and then followed up in many studies later [48, 49]. However, it is found that very few simulations over the years find a successful explosion when studied in 1 dimension, i.e. only with the radial coordinate. This will be discussed in detail later.

If L_{ν_e} is the luminosity of electron type neutrinos at the surface of their neutrinosphere with radius R_ν and assuming, for simplicity, the luminosity of the electron type antineutrinos on their neutrinosphere surface to have the same value as L_{ν_e}, the rate of heating of matter at a radial distance R_m for $R_m \gg R_\nu$ is given by Bethe and Wilson [46]

$$\frac{dE_+}{dt} = \frac{1}{4\pi R_m^2} K(T_\nu)(Y_n + Y_p)L_{\nu_e} , \qquad (2.39)$$

where Y_n and Y_p are the fractions of protons and neutrons, respectively, neglecting the small heating on the nuclei to simplify ($Y_N = Y_n + Y_p$). The reactions involved in the heating are $n + \nu_e \rightarrow p + e^-$ and $p + \bar{\nu}_e \rightarrow n + e^+$. The variable $K(T_\nu)$ is the neutrino absorption coefficient in $cm^2\ g^{-1}$. As the stalled shock remains normally at distances 100–200 km from the center of the core, the distance R_m is indeed substantially larger than the neutrinosphere radii. The absorption coefficient K is related to the neutrino/antineutrino capture cross section σ by

$$K = N_{Av} < \sigma >= 9 \times 10^{-44} N_{Av} < E_\nu^2 > cm^2 g^{-1} , \qquad (2.40)$$

where N_{Av} is the Avogadro number and the cross section, depending on the square of the $\nu_e/\bar{\nu}_e$ energy is averaged over the spectrum of the neutrinos/antineutrinos coming from their neutrinosphere. For $< E_\nu^2 >$ we use $\sim 21 T_\nu^2$ with the temperature in MeV, as mentioned by Bethe and Wilson [46], though the spectrum is supposed to have less higher energy neutrinos compared to the Fermi or the Boltzmann distribution, as the neutrino mean free path is shorter for higher energies.

There is also deposition of energy on the matter by neutrinos scattering from the electrons. The rate of heating by the scattering is roughly equal to $(\frac{T_m}{T_\nu})\frac{dE_+}{dt}$ [46] where dE_+/dt is the heating rate through the neutrino/antineutrino absorption given by Eq. (2.39). For typical values of the two temperatures, it is roughly a 20% effect. However, we shall not include the contribution of this scattering effect in the following discussion and take $Y_N = 1.0$ neglecting the nuclei. Finally with the temperature in units of 4 MeV, L_{ν_e} in 10^{52} ergs and the radial distance in 100 kms, one gets [48]

$$\frac{dE_+}{dt} = 1.54 \times 10^{20} L_{\nu_e} \left(\frac{100\ km}{R_m}\right)^2 \left(\frac{T_{\nu_e}}{4MeV}\right)^2 \left[\frac{erg}{g\ s}\right] . \qquad (2.41)$$

Now the matter inside the shock front has the nuclei dissociated into protons and neutrons. But the matter in front of the shock and the matter falling through the shock are in the form of nuclei from iron-type to ^{16}O. The neutrino capture cross section on them is much smaller than the ones on the nucleons. Even when they are broken into alpha particles, the cross section is still very small and negligible. So for most matter outside the shock, the value of Y_N is considerably smaller than one and that should be taken into account.

The matter in the shock region also gets cooled by the emission of neutrinos and antineutrinos from the matter at a temperature T_m. This again can be written in terms of $K(T_m)$ for the electron and positron capture reactions that create the neutrinos and antineutrinos. Also the energy per unit volume of a "blackbody" neutrino gas is given by acT_m^4. The assumption that the neutrino gas in the region of the stalled shock can be treated as a "blackbody" is a reasonable one, as shown by Bethe and Wilson [46] and hence the rate of cooling of the matter at temperature T_m is

$$\frac{dE_-}{dt} = K(T_m)acT_m^4 ,$$

(2.42)

where $a = 0.60 \times 10^{26}$ ergs cm^{-3} Mev^{-4} [46] and 'c' is the velocity of light, as earlier. Then putting the value of the constants, one eventually gets the rate of cooling as [48]

$$\frac{dE_-}{dt} = 1.40 \times 10^{20} \left(\frac{T_m}{2MeV}\right)^6 \left[\frac{erg}{g\,s}\right] ,$$

(2.43)

and the net rate of heating is given by

$$\frac{dE_0}{dt} = \frac{dE_+}{dt} - \frac{dE_-}{dt} .$$

(2.44)

Actually one can write the luminosity L_{ν_e} in Eq. (2.39) as proportional to T_ν^4 and show a heating rate proportional to the sixth power of the neutrino temperature. From that one can conclude that to receive a net positive heating, the matter temperature should be lower than a maximum value, determined by the temperature of the neutrinosphere [46].

In the 1980s and 1990s, a number of 1-dimensional calculations/simulations were done involving the radial coordinate assuming spherical symmetry. However slowly it became clear almost all such simulations do not lead to explosions. The role of convection involving the angular coordinates, mentioned in Sect. 2.4, and different instabilities that help the transfer of energy to the outer parts were recognized. Let us write down the basic equations of hydrodynamics that one uses initially without effects from general relativity. They are the usual equations of conservation of mass, momentum, and energy and written as [48]

$$\frac{d\rho}{dt} = -\rho\nabla.\mathbf{v} ,$$

(2.45)

$$\rho\frac{d\mathbf{v}}{dt} = -\rho\nabla\phi - \nabla P ,$$

(2.46)

and

$$\rho\frac{d\epsilon}{dt} = -P\nabla.\mathbf{v} + \frac{dE_+}{dt} - \frac{dE_-}{dt} ,$$

(2.47)

Here ρ is the mass density, $\mathbf{v}$ is the fluid velocity, Φ is the potential due to gravity. The isotropic pressure is given by P and the specific internal energy is denoted by

ϵ. The heating rate of the matter $\frac{dE_+}{dt}$ and the cooling rate $\frac{dE_-}{dt}$ are as given in Eqs. (2.41) and (2.43), respectively.

Next, we briefly describe, following O'Connor and Ott [50, 51], the construction of spherically symmetric general relativistic Eulerian hydrodynamic code known as GR1D, for the matter with three species of neutrino treated in a leakage/heating scheme. In this code, the hydrodynamical and 3+1 spacetime evolution equations are worked out in the radial gauge polar slicing coordinate system [52]. This uses the units $G = c = M_\odot = 1$ and spacelike signature $(-,+,+,+)$. The line element is written as, [50]

$$ds^2 = -\alpha^2(r, t)dt^2 + X^2(r, t)dr^2 + r^2 d\Omega^2 , \qquad (2.48)$$

where $\alpha(r, t) = exp(\Phi(r, t))$ with $\Phi(r, t))$ being the metric potential and $X(r, t) = [1 - 2m(r)/r]^{-1/2}$ with $m(r)$ as the enclosed gravitational mass. The stress-energy tensor of the ideal fluid and matter current density are expressed as

$$T^{\mu\nu} = \rho h u^\mu u^\nu + g^{\mu\nu} P , \qquad (2.49)$$

along with

$$J^\mu = \rho u^\mu , \qquad (2.50)$$

where u^μ is the 4-velocity of the fluid and can be expressed as $u^\rho = (\Gamma/\alpha, \Gamma v^r, 0, 0)$ where $\Gamma = (1 - v^2)^{-1/2}$ and $v = X v^r$ is the physical velocity. Here ρ is the matter density and P the fluid pressure. The specific enthalpy (h) is related to specific internal energy (ϵ) by $h = 1 + \epsilon + P/\rho$. The Hamiltonian and momentum constraint equations are [50, 53]

$$m(r) = 4\pi \int_0^r (\rho h \Gamma^2 - P + \tau_m^\nu) r'^2 dr' , \qquad (2.51)$$

$$\Phi(r, t) = \int_0^r X^2 \left[\frac{m(r', t)}{r'^2} + 4\pi r'(\rho h X^2 u^r u^r + P + \tau_\Phi^\nu) \right] dr' + \Phi_0 . \qquad (2.52)$$

The effects of trapped neutrinos are included in the τ_m^ν, τ_Φ^ν terms.

The quantity Φ_0 is evaluated by matching the metric on the star's surface to the Schwarzschild metric. The evolution equations of the fluid come from expanding the local fluid rest-frame conservation equations,

$$\nabla_\mu T^{\mu\nu} = 0 , \qquad (2.53)$$

$$\nabla_\mu J^\mu = 0 . \qquad (2.54)$$

In this code, the flux-conservative Valencia formulation modified for spherically symmetric flows is used [54, 55].

The evolution equations become [50]

$$\partial_t \mathbf{U} + \frac{1}{r^2} \partial_r \left(\frac{\alpha r^2}{X} \mathbf{F} \right) = \mathbf{S} , \tag{2.55}$$

with $\mathbf{U} = (D, DY_e, S^r, \tau)$ are conserved variables which are functions of the variables ρ, Y_e, ϵ, v, and P and

$$\mathbf{U} = \begin{pmatrix} D \\ DY_e \\ S^r \\ \tau \end{pmatrix} = \begin{pmatrix} X\rho\Gamma \\ X\rho\Gamma Y_e \\ \rho h \Gamma^2 v \\ \rho h \Gamma^2 - P - D \end{pmatrix} . \tag{2.56}$$

$\mathbf{F}$ is the flux vector,

$$\mathbf{F} = (Dv, DY_e v, S^r v + P, S^r - Dv) . \tag{2.57}$$

S is the source terms containing all the gravitational and matter interaction sources,

$$S = \left[0, \; R^v_{Y_e}, \; (S^r v - \tau - D)\alpha X \left(8\pi r P + \frac{m}{r^2} \right) + \alpha P X \frac{m}{r^2} \right.$$
$$\left. + \frac{2\alpha P}{Xr} + Q^{v,E}_{S^r} + Q^{v,M}_{S^r}, \; Q^{v,E}_{\tau} + Q^{v,M}_{\tau} \right] . \tag{2.58}$$

The Rs and Ss are the neutrino source and sinks coming from the neutrino leakage scheme and contributions from neutrino pressure [50]. The Eq. (2.55) is discretized in space using a finite volume scheme and specific numerical methods used to reconstruct the state variables at the cell interfaces. In this spherically symmetric situation, rotation can be included by introducing an angular velocity Ω on the spherical shells and an averaged centrifugal force in the radial motion equation.

The GR1D code includes the microphysical EoS and an approximate neutrino scheme for the pre- and postbounce phases. A computationally expensive Boltzmann transport is replaced by an efficient scheme for neutrinos here. Furthermore, the approximate treatment of neutrinos in the GR1D was in agreement with the salient features of 1D radiation hydrodynamics simulations using Boltzmann neutrino transport [51]. In the prebounce phase, neutrinos are produced due to the capture of electrons on nuclei and free protons. As a result, the electron fraction and mass of the homologous core decrease. It was shown that the electron fraction Y_e could be effectively parametrized as a function of density [56]. This was implemented in the GR1D code. However, this simple parametrization does not hold good in the postbounce phase due to multiple effects of deleptonization, neutrino cooling and heating. So, a three-flavor, energy averaged neutrino leakage scheme was adopted to capture those effects [57, 58]. In this code three types of neutrino species were denoted by ν_e, $\bar{\nu}_e$, $\nu_x (= \nu_\mu, \bar{\nu}_\mu, \nu_\tau, \bar{\nu}_\tau)$ [50, 51]. The leakage scheme exploited in the GR1D code gives approximate energy emission rates. To obtain an explosion in the spherically symmetric model, an artificial neutrino heating in the post shock region was considered in the model. It was based on a parametrized charged current scheme [59]. The local neutrino heating rate for the neutrino species ν_i is given by,

$$Q_{\nu_i}^{heat}(r) = f_{heat} \frac{L_{\nu_i}(r)}{4\pi r^2} \sigma_{heat, \nu_i} \frac{\rho}{m_u} X_i \left\langle \frac{1}{F_{\nu_i}} \right\rangle e^{-2\tau_{\nu_i}} , \qquad (2.59)$$

where the scale factor f_{heat} takes values higher than 1 to add an extra neutrino heating for CCSN explosions [50]. Various quantities in Eq. (2.59) are defined in [51].

Ott and O'Connor exhaustively investigated the black hole formation in failed CCSN simulations using different progenitor models, which were the outcome of various stellar evolution studies as well as using non-relativistic Lattimer-Swesty (LS) and relativistic HShen nuclear EoS [50]. This led to the systematic study of a large parameter space that determines the fate of massive stars in gravitational core collapse. It has been noted that the results obtained in CCSN simulations using the simplified treatment of neutrino leakage and heating in the GR1D were qualitatively similar to the results obtained from one dimensional simulations with the Boltzmann neutrino transport by other groups [60, 61]. Furthermore, neglecting the effects of multi-dimensional dynamics on neutrino transport, the scale factor f_{heat} is increased above the default value of unity to get the required neutrino heating efficiency for CCSN explosions. The important quantity here is the time averaged heating efficiency of the critical model as defined in Ref. [50]. This determines how much of the neutrino luminosity would be deposited on average for a CCSN explosion. They obtained a value of 0.13 for a 15 $M_\odot$ zero age main sequence solar metallicity progenitor of Woosley and Weaver. A comparison of this value with the value 0.07 for the same progenitor in the neutrino-driven explosion in 2D [62] clearly demonstrates the role of dimensionality on neutrino heating.

Another important input in CCSN simulations is the EoS of matter over a wide range of temperatures ($\sim$ 0–150 MeV), densities ($10^4 - 10^{15}$ g/cm^3) and proton fraction (0–0.6). The EoS controls many aspects of CCSN such as the neutrino-matter interaction, the core bounce, the protoneutron star (PNS), the collapse of the PNS to a black hole or not, the composition of the ejected material, and the gravitational wave signal. Many studies were performed on the role of EoSs on those aspects of CCSN [50, 63–72]. Investigations of CCSN simulations with EoSs involving exotic phases of matter (hyperons and quarks) are notable among all those studies [69, 73]. The main question is whether a phase transition from nuclear matter to hyperon or quark matter could lead to a successful CCSN explosion.

The CCSN simulations of 20–40 $M_\odot$ Woosley and Heger (WH) progenitor models using the EoS involving Λ hyperons known as Banik, Hempel, and Bandyopadhyay (BHB) $\Lambda\phi$ EoS compatible with the observations of 2 $M_\odot$ neutron stars were carried out by Char et al. [69, 74]. The results were also compared with those of the HShen Λ EoS which does not lead to a 2 $M_\odot$ cold neutron star [75].

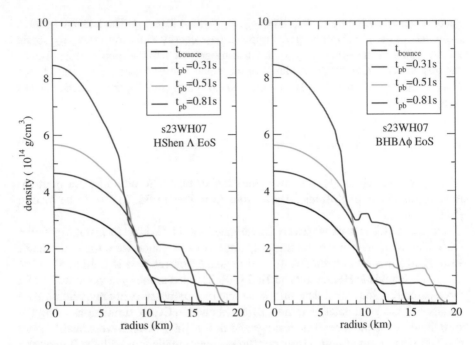

Fig. 2.4 Density profiles of the PNS are exhibited as a function of radius at the core bounce, and postbounce time (t_{pb}) 0.31, 0.51 and 0.81 s for the progenitor model s23WH07 and HShen Λ EoS (left panel) and BHBH$\Lambda\phi$ EoS (right panel)

Various thermodynamic variables are discussed below for both EoSs at different time snap-shots.

The density profiles as a function of radius are plotted for s23HW07 at the bounce as well as for postbounce times, $t_{pb} = 0.31$, 0.51 and 0.81 s in Fig. 2.4. The left panel of the figure represents the HShen Λ EoS and the right panel relates to the results of the BHB$\Lambda\phi$ EoS. At the bounce, the central density (ρ_c) of the PNS in both cases is 30–40% above the normal nuclear matter density. Their evolution with time follows the same pattern and the central density increases at later times. However, the density falls well below normal nuclear matter density above 14 km. The high central density is responsible for the population of Λ hyperons in the core of the PNS.

The temperature profiles as a function of radius are shown for s23WH07 with the HShen Λ hyperon (left panel) and BHB$\Lambda\phi$ (right panel) EoSs at different times in Fig. 2.5. The peaks of temperature profiles are more than 10 km away from the center of the PNS in both cases after the core bounce. Later those move toward the center. The peak temperature around 8 km in case of the BHB$\Lambda\phi$ EOS is somewhat higher than that of the HShen Λ EoS at 0.81 s. This high temperature would result in thermally produced Λ hyperons move away from the center of the PNS.

The PNS compositions as a function of radius at two different postbounce times for two progenitor models s23WH07 (left panel) and s40WH07 (right panel) with

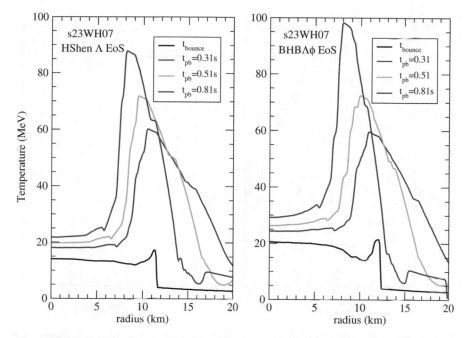

Fig. 2.5 Temperature profiles of the PNS are depicted as a function of radius at the core bounce, and postbounce time (t_{pb}) 0.31, 0.51 and 0.81 s for the progenitor model s23WH07 and HShen Λ EoS (left panel) and BHB$\Lambda\phi$ EoS (right panel)

BHBH$\Lambda\phi$ EoS are shown in Fig. 2.6. It is seen that the population of Λs increases with time. Thermally produced Λs are more abundant around 8 km at a later time of 0.51 s for the BHB$\Lambda\phi$ EoS due to a higher peak temperature.

It is worth noting that the central value of Λ fraction is a high density effect whereas the off center Λs are populated thermally. Finally, for both EoSs and the progenitor model discussed here, the PNS collapses to a black hole around ~0.88 s.

So far it is found that simulations in 1-dimensional CCSN models might lead to accretion driven black holes in failed CCSNs. It was long debated whether exotic phases of matter might make the PNS metastable after a successful CCSN and the PNS would collapse to a low mass black hole during the long duration evolution when the thermal support decreases and the deleptonization takes place. This kind of scenario was envisaged for the non-observation of a compact object in SN1987A and addressed in CCSN simulations [69, 76–79]. This study was performed in the GR1D model by increasing the neutrino heating scale factor to $f_{heat} = 1.5$ for s20WH07 with the BHB$\Lambda\phi$ EoS [69]. The shock radius is shown as a function of postbounce time in the left panel of Fig. 2.7. In the right panel, the evolution of the PNS gravitational mass is exhibited with postbounce time. For neutrino scale factor $f_{heat} = 1$ (black), the shock radius recedes and finally, the PNS collapses into a black hole whereas it is observed that the shock radius increases with time in case of $f_{heat} = 1.5$ (red), after a successful supernova explosion. The PNS is found to be

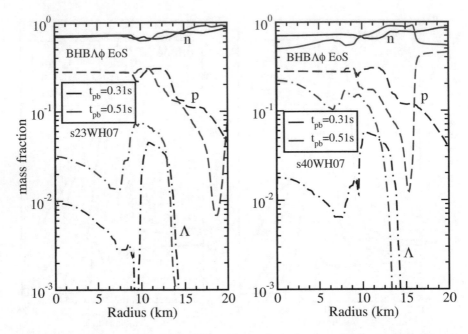

Fig. 2.6 Mass fraction profiles of the PNS are depicted as a function of radius at the postbounce time (t_{pb}) 0.31 and 0.51 s for the progenitor model s23WH07 (left panel) and s40WH07 (right panel) for the BHBH$\Lambda\phi$ EoS. This is taken from Ref. [69] and reproduced by permission of the American Astronomical Society (AAS). https://doi.org/10.1088/0004-637X/809/2/116

stable in long term evolution till 4 s. The PNS might evolve to a cold neutron star as no onset of the metastability in the PNS due to the loss of thermal support and neutrino pressure, during the cooling phase over a few seconds is found.

It is worth mentioning here that a second shock arising out of a phase transition to quark matter in the early postbounce evolution of CCSN of low and intermediate-mass progenitor stars was reported in the simulations of Sagert et al. [73]. There the second shock eventually triggered a delayed supernovae explosion. A clear imprint of the phase transition should have observable consequences.

2.7 Multi-Dimensional Hydrodynamic Simulations and the Present Scenario

There is an overall agreement today among physicists, working on core collapse supernova modelling that 1-dimensional spherically symmetric realistic simulations do not lead to explosions. The only exception may be some stars of masses 8–10 $M_\odot$ which develop instead of an iron core mass, a core of O-Ne-Mg with a thin carbon shell plus a loosely bound He shell outside of it and give rise to either the prompt

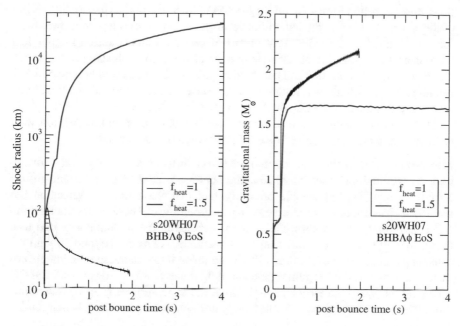

Fig. 2.7 Shock radius (left panel) and PNS gravitational mass (right) for f_{heat} =1 (black) and 1.5 (red) with the progenitor s20WS07 and BHB$\Lambda\phi$ EoS This is taken from Ref.[69] and reproduced by permission of the American Astronomical Society (AAS). https://doi.org/10.1088/0004-637X/809/2/116

explosions or delayed explosions by neutrino heating in 1-dimensional simulations [80, 81]. However, this may lead to a faint type II supernova with energies a bit more than 10^{50} ergs [82]. Before going to the details of multi-dimensional simulations, one needs to highlight some of the essential physics items that are important. They are:

Convection The importance of convection is long recognized [1, 83, 84]. The proposals for the location and the duration of the convection process helping the energy transfer, changed over the last few decades and one presently envisages two regions : the first one is inside the PNS below the neutrinospheres, caused by a negative lepton number gradient. The second one is in the gain region immediately below the shock, where there is more neutrino heating than cooling, and this is driven by a negative entropy gradient [85]. The convection inside the PNS starts 30–40 ms after bounce and results in higher luminosities of the $\bar{\nu}_e$ by $\sim$ 15% and even larger for ν_x, but with reduced average energies. It seems to depend on the nuclear symmetry energy. The outer convection has many effects like transporting energy from near the PNS zone to the stalled shock. It changes the accretion pattern as well as the accretion rate. Accretion gets directed in narrow channels of the material moving downward with hot, lower density bubbles around them. In Fig. 2.3 we see the convection circles inside the PNS as well as in the outer ones, in the lower left panel.

Standing Accretion Shock Instability (SASI) [86, 87] This is a generic instability of the accretion shock to the non-radial deformation and shows largest growth rates for $l = 1$ and $l = 2$ modes. That means, if one does an expansion in spherical harmonics, the dipole and quadrupole terms cause a bipolar sloshing of the shock along with strong expansion and contraction pulses [28]. The rapid growth of SASI is caused by perturbations resulting from the onset of convection on the postshock matter and this takes place also on the convectively stable matter.

The SASI and neutrino-driven convections are often seen in competition with each other and the accretion rate determining which one dominates.

Many-Body Corrections to Neutrino-Matter Interaction Many-body corrections to NC and CC neutrino-matter interaction are studied by different groups over the years [82, 88]. People presently feel that these have an effect on the neutrino-driven mechanism of core collapse supernovae. These corrections on the NC neutrino-nucleon scattering lead to an increase in the ν_x luminosity due to a decrease in the associated scattering cross section for densities beyond 10^{12} g/cm^3 causing a compression of the core [49]. They result in the increase of the neutrino-matter heating rates in the gain region and help the explosion. However for the CC involving the $\nu_e/\bar{\nu}_e$-matter interaction, since the superallowed (Fermi) transitions dominate the cross sections, the many-body correction induced effect is marginal.

Progenitor Model It appears that the density profile of the progenitor stars has a strong effect on the possibility of explosion. The progenitor models with very steep outer density profiles which give rise to a fast decreasing postbounce accretion rate, have a better chance of explosion. They even result in explosions in a few 1D simulations as mentioned earlier, but not in the others [89].

Turbulence Some of the recent simulations have brought attention to turbulence which effectively stiffens the EOS of the material behind the shock and gives stronger pressure for a given density there. This causes the shock to get pushed outward making the gain region bigger [90].

There are coordinated international efforts involving large multi-dimensional simulations that are able to give detailed results today. Among these groups the Princeton supernova group uses the "state-of-the-art" code FORNAX [91] as discussed by Burrows and Vartanyan [89] and present detailed results for progenitor masses 9–27 M$_\odot$. Out of the 18 cases with different stellar masses reported only two do not explode and the others have explosion energies varying from 0.09 to 2.3 times 10^{51} ergs with run times of 1.04 to 4.66 seconds after the bounce. All simulations were evolved in 2D axisymmetry with 1024 radial cells and 128(θ) cells [89]. These 2D simulations were performed to map some of the parameter space and give foresight into the essential physics. With the huge increase of computer power over the decades, today it is possible to run multiple 3D simulations per year. Some of the 3D simulations by the Princeton supernova group were performed with 678 (radius) $\times$ 256 (θ) $\times$ 512 (ϕ) grid [89, 92]. It is seen by the group that when a 2D model explodes, the more realistic 3D counterpart does explode too.

The results and the discussions of the inherent physics of the 2D (and the early 3D) simulations of the "Garching" group can be found in the papers of Janka et al. [5, 93]. An exhaustively detailed description of 3D as well as 2D simulations by different groups in recent years is presented by Müller [94].

Let us briefly outline some of the present-day multi-dimensional simulation results. Not all of these may turn out to be correct, but they are indicative of the physics issues involved. If one decomposes the radius of the stalled shock in spherical harmonics in solid angle, the onset of the explosion is reflected in the monopolar instability while the dipole is also seen to become unstable simultaneously. So even for a non-rotating star, the explosion selects an axis more-or-less randomly and the extent of the asymmetry seems to depend on the mass, with the heavier mass cores exploding later and the turbulence making them more chaotic. In such cases, one can describe the system by distribution functions of explosion energies, directions, times, residual masses, etc. of the star but the details of these functions are difficult to figure out for a particular star.

The breaking of symmetry in going from 1D to multi-dimensional simulations is important as it allows accretion. Simultaneous accretion in one direction and explosion in another along with the turbulent stress is often seen in these simulations. In the progenitor models, the role of aspherical perturbations in maintaining the turbulent convection behind the stalled shock is an area where many present and future codes plan to concentrate on [89]. Also, the convection in the region of the protoneutron star as mentioned earlier, is seen to play a role in accelerating energy loss through the neutrino and electron loss in PNS leading to core shrinkage, thus helping the explosion. The other important aspect appears to be the mass density structure just before the collapse and its effect on the subsequent evolution. It appears that steep density profiles for smaller stellar masses end in a faster explosion. More full 3D results of different groups in the near future may come to an agreement on these aspects.

Roughly about 1% of core collapse supernovae are termed "hypernovae" which are believed to be connected to the emission of a long-soft Gamma Ray Burst (GRB), with explosion energies in the range of 10^{52} ergs. These are too energetic to be powered by the neutrino mechanism. They are believed to be powered by magnetohydrodynamic jets that use the large spin kinetic energy of the fast rotating protoneutron star. This area is also being studied now extensively.

To conclude, future comparisons of results of the multi-dimensional simulations of different groups would lead to a better understanding of the final stages of the explosion and hopefully, solve a century-old problem.

2.8   The Supernova SN1987A

The supernova 1987A, already mentioned in Sect. 1.1, has for many reasons turned out to be one of the most studied objects in the history of astronomy. We shall describe some of these aspects in this section. SN 1987A has been classified as a peculiar type II supernova because of its long rising light curve and the persistence of H1 lines in its spectra. Also because its progenitor is a blue supergiant (BSG) and not a red supergiant (RSG). Since then some more type II-peculiar have been observed like SNe 2000cb, 2006V, 2006au, 2009E, 2009mw and their explosions have been modelled generating explosions with a range of energies and ^{56}Ni masses [95]. After the observation of the neutrinos during 1987A, the details of which we describe later in this section, people expected to witness the birth of a neutron star. However that turned illusive. Owing to the dust and the ring around its supernova remnant, radio and X-ray searches for a neutron star gave negative results earlier. However after more than thirty years, recently Cigan et al. [96] show high-angular-resolution images of dust and molecules in the SN1987A remnant from the Atacama Large Millimeter/sub-millimeter Array (ALMA) facility indicating the presence of a compact source. The observed infrared excess is believed to be most probably from an embedded, cooling neutron star NS1987A [97]. As a result, there is a lot of interest to learn more about the newborn neutron star through the present and future observations.

The neutrinos detected by the water Cerenkov detectors IMB [98] and K II [99], as mentioned in Sect. 1.1 created a lot of excitement. The two processes that can take place on the protons and electrons of the water in the detectors are the capture reaction $\bar{\nu}_e + p \rightarrow n + e^+$ and the scattering reaction $\nu_e + e^- \rightarrow \nu_e + e^-$. The details of the times and energies of the observed events are given in Table 2.1. Other possibilities of scattering events of electrons with the electron antineutrinos or μ or τ type neutrino/antineutrinos have smaller cross sections and are not considered here. Also as discussed below, the antineutrino capture events have much larger cross sections and much more probable than the scattering of electrons. Hence we take the events all as $\bar{\nu}_e$ capture on protons. In Table 2.1 we present the time, the energy deposited on the positron and the angle θ_e between the directions of the positron emitted and the incoming $\bar{\nu}_e$.

It should be mentioned here that two other scintillation detectors, Baksan [100] in Soviet Union and Mont Blanc in Europe [101] also reported observation of neutrinos. But the signals in Mont Blanc occurred 4.7 hours before the simultaneous observations in K II and IMB and was considered controversial. Similarly, the Baksan events are also not included in most analyses.

In case of the capture of the electron antineutrino, the energy of the supernova $\bar{\nu}_e$, $E_{\bar{\nu}_e}$ for each observed event can be expressed as [102]

$$E_{\bar{\nu}_e} = \frac{E_e + \Delta + (\Delta^2 - m_e^2)/2m_p}{1 - (E_e - p_e cos\theta_e)/m_p} , \qquad (2.60)$$

where $\Delta = m_n - m_p = 1.293\,\text{MeV}$, $E_e(p_e)$ are the energy(momentum) of the positron. For the scattering on electrons, one can also relate the supernova neutrino energy to the electron energy by a similar equation. But the capture event has a much

Table 2.1 The relative times of the events, their energies and angles observed at IMB and Kamiokande II [98, 99, 103]

Observed at	Event no	Time (s)	Positron energy (MeV)	Angle θ_e (degree)
IMB	1	0	38 ± 7	80 ± 10
	2	0.412	37 ± 7	44 ± 15
	3	0.650	28 ± 6	56 ± 20
	4	1.141	39 ± 7	65 ± 20
	5	1.562	36 ± 9	33 ± 15
	6	2.684	36 ± 6	52 ± 10
	7	5.010	19 ± 5	42 ± 20
	8	5.582	22 ± 5	104 ± 20
K II	1	0	20.0 ± 2.9	18 ± 18
	2	0.107	13.5 ± 3.2	40 ± 27
	3	0.303	7.5 ± 2.0	108 ± 32
	4	0.324	9.2 ± 2.7	70 ± 30
	5	0.507	12.8 ± 2.9	135 ± 23
	6	0.686	6.3 ± 1.7	68 ± 77
	7	1.541	35.4 ± 8.0	32 ± 16
	8	1.728	21.0 ± 4.2	30 ± 18
	9	1.915	19.8 ± 3.2	38 ± 22
	10	9.219	8.6 ± 2.7	122 ± 30
	11	10.433	13.0 ± 2.6	49 ± 26
	12	12.439	8.9 ± 1.9	91 ± 39

larger cross section than the scattering event as given by

$$\sigma(\bar{\nu}_e - p) = 8.9 \times 10^{-42} \text{cm}^2 \left(\frac{E_e}{10\,\text{MeV}}\right)^2, \tag{2.61}$$

whereas

$$\sigma(\nu_e - e) = 9.4 \times 10^{-44} \text{cm}^2 \left(\frac{E_\nu}{10\,\text{MeV}}\right). \tag{2.62}$$

Theoretical models predict the flux of ν_es somewhat higher than the flux of $\bar{\nu}_e$s and with the fact that each water molecule have 2 protons and 10 electrons, one expects the number of scattering events to be 0.06–0.12 times smaller than the number of capture events for typical 10 MeV $\nu_e/\bar{\nu}_e$s. Only those events that deposit energies higher than 5 MeV could be observed by the Cerenkov detectors and so one did not consider the scattering events on protons. The IMB and K II detectors had 5000 and 3000 tons of water of which 3300 and 2140 tons were the fiducial masses, respectively. The capture reaction is approximately isotropic with $\frac{d\sigma}{d\cos\theta} \sim [1 + 0.11(1 - \cos\theta)]$ where the forward and backward directions differ in the differential cross section by 22%. The scattering events, however, are peaked in the forward direction.

These observations gave very important information about two different aspects: (a) about the conditions in the supernova and (b) about the properties of the particle, neutrino/antineutrino.

First, we discuss how one can estimate the total energy emitted in electron type antineutrinos, their average energy, and the source temperature [104]. One assumes that the energy distribution of the $\bar{\nu}_e$s does not change with time and is given by a Fermi distribution having zero chemical potential, $f_F(E)$ with constant temperature "T." The temperature here is expressed in energy units. The average energy of the antineutrinos, $< E_{\bar{\nu}} >_s$ at the source is given by

$$< E_{\bar{\nu}} >_s = \frac{\int_0^\infty E_{\bar{\nu}}^3 f_F(E_{\bar{\nu}}) dE_{\bar{\nu}}}{\int_0^\infty E_{\bar{\nu}}^2 f_F(E_{\bar{\nu}}) dE_{\bar{\nu}}} = \frac{F_3(0)}{F_2(0)} T = 3.15\, T \,, \tag{2.63}$$

where the function $F_k(\eta)$ is the standard Fermi integral given by

$$F_k(\eta) = \int_0^\infty \frac{\epsilon^k d\epsilon}{1 + exp(\epsilon - \eta)} \,. \tag{2.64}$$

The average energy of the antineutrinos at the detector, $< E_{\bar{\nu}} >_d$ can be written as [105]

$$< E_{\bar{\nu}} >_d = \frac{\sum_{i=1}^{N_d} E_{\bar{\nu}i}}{N_d} = \frac{\int_H^\infty E_{\bar{\nu}}^5 f_F(E_{\bar{\nu}}) W(E_{\bar{\nu}}) dE_{\bar{\nu}}}{\int_H^\infty E_{\bar{\nu}}^4 f_F(E_{\bar{\nu}}) W(E_{\bar{\nu}}) dE_{\bar{\nu}}} = T \frac{G_5(H/T)}{G_4(H/T)} \,, \tag{2.65}$$

where $E_{\bar{\nu}_i}$ (i=1,2, ... N_d) are the energies of the detected neutrinos. $W(E_{\bar{\nu}})$ is the detector efficiency which is a known function and H is the threshold energy of the detector. $G_n(x)$ is the truncated Fermi integral modified by the threshold. The Eq. (2.65) uses the fact that the antineutrino capture cross section is proportional to the square of their energy and can be solved by an iterative procedure to find the temperature T. Then Eq. (2.63) will give the average energy of the antineutrinos emitted from the supernova. The total energy in $\bar{\nu}_e$ from SN 1987A can then be expressed as

$$E_{\bar{\nu}_e}^{Tot} = 0.77 \times 10^{52} \left(\frac{D}{50 kpc}\right)^2 \left(\frac{1}{M}\right) \left[\frac{F_3(0)G_5(H/T)}{G_4^2(H/T)}\right]$$

$$\times \left[\frac{10\,\text{MeV}}{< E_{\bar{\nu}} >_d}\right] N_d\ ergs \,, \tag{2.66}$$

where D is the distance of the LMC and M is the fiducial mass in kton. The results for such a calculation gives [104] the average energy, the temperature and the total $\bar{\nu}_e$ energy for the K II events as 8.57 (± 0.82) MeV, 2.72 (± 0.26) MeV and 75 ($^{+36}_{-14}$)

$\times 10^{51}$ ergs, respectively. For the IMB events the corresponding values are 14.1 ($\pm$ 1.9) MeV, 4.48 ($\pm$ 0.74) MeV and 34 ($^{+41}_{-14}$) $\times 10^{51}$ ergs, respectively. There are other calculations also available for these quantities and they all give valuable information about the supernova itself.

Finally, we discuss how the data of neutrinos from SN1987A can put limits on the mass of neutrinos and other neutrino properties. For this, we focus on the two following aspects: (i) All the observed neutrino events were detected in a time span of 10–12 seconds (ii) All the neutrinos have energies less than, say 60 MeV.

Today after the discovery of the oscillations of atmospheric and solar neutrinos, one knows there are two non-zero mass squared differences for the three active neutrinos and the flavors mix as neutrinos move in space. But in 1987 it was useful to set limits on neutrino masses [103]. For that one notes that the difference in the time at the detector (t_d) and the time of emission from the source (t_s) for a neutrino with mass m_ν and energy E_ν is given by

$$\Delta t = t_d - t_s = \frac{D}{c}\left[1 + m_\nu^2/(2E_\nu^2)\right], \qquad (2.67)$$

where, as earlier D is the distance of the source. So the delays due to the mass m_ν is given by $\frac{1}{2}\frac{D}{c}\frac{m_\nu^2}{E_\nu^2}$. So for two neutrinos with energies E_1 and E_2, one finds the absolute value of the difference of their time delays is [103]

$$\bar{\Delta} = |\Delta t_d - \Delta_s| = \frac{D}{2c}m_\nu^2\frac{|E_2^2 - E_1^2|}{E_2^2 E_1^2}. \qquad (2.68)$$

Experimentally Δt_d is less than 12 seconds. To illustrate, if we take the K II events 3 and 11, then using $E_1 = 7.5$ MeV, $E_2 = 13.0$ MeV, $\Delta t_d = 10.1s$ and assuming $\Delta t_s \le 10s$, one gets $m_\nu < 18$ eV. Detailed analysis finds the upper limit on mass to be in the range of 19–30 eV.

Another quantity that can be easily obtained is the lower limit on neutrino lifetime if one assumes the neutrinos/antineutrinos to be unstable with a very large lifetime. Assuming the neutrino lifetime in the rest frame as t_{ν_e} and including the effect of time dilation for neutrinos of mass m_ν and energy E_ν, one can write [103] for neutrinos traveling the distance D without decaying

$$\left(\frac{E_\nu}{m_\nu}\right) t_{\nu_e} \ge \frac{D}{c}, \qquad (2.69)$$

$$t_{\nu_e} \ge \left[\frac{m_\nu/1eV}{E_\nu/10\,\mathrm{MeV}}\right] \times 5 \times 10^5 \text{ s}. \qquad (2.70)$$

This says that for a neutrino of mass 1 eV and energy 10 MeV the lifetime is greater than 5×10^5 seconds. This itself was useful in ruling out neutrino decay as a possibility to explain the solar neutrino puzzle.

Using similar arguments one got upper limit (consistent with zero) on neutrino charge, neutrino magnetic moment, the strength of right-handed weak interaction etc. For a detailed discussion, we refer to the book by Mohapatra and Pal [103]. Finally we stress that though the number of detected 1987A neutrino events is small, it is often useful for verifying or constraining any new theoretical idea in neutrino physics, as in neutrino physics beyond the standard model. This has resulted in hundreds of research papers over the years.

2.9   Detection of Neutrinos from Future Supernova Events

The capabilities for detecting neutrinos from supernovae have increased by orders of magnitude in the last thirty four years after SN1987A. As the core collapse supernova events in galaxies like the Milky Way is estimated as a few per century [106], there is a good possibility of observing a core collapse supernova by us in the next few decades. The most likely distance of such a galactic supernova is estimated as 12–15 kpc [107]. That will provide us a lot of information about the properties of neutrinos as well as about supernova mechanisms [85, 106].

Before we go into the details of the detection processes available today, we briefly describe the knowledge we have today about neutrino mass and neutrino oscillations, acquired during and after the analysis of the atmospheric neutrinos at the Superkamiokande (SK) detector along with their zenith angle dependence followed by the observation of the solar neutrinos by the Sudbury Neutrino Observatory (SNO) through electron scattering and charged/neutral current dissociation of heavy water. Of course the deficit of the detected solar neutrinos observed by the Chlorine and Gallium radiochemical detectors, compared to their theoretical estimates, known as the solar neutrino puzzle, existed for decades before that [103].

The flavor eigenstates $|\nu_\alpha >$ created inside the supernova can be expressed in terms of a linear superposition of the mass eigenstates $|\nu_i >$ as $|\nu_\alpha >= \Sigma_i U_{\alpha i}|\nu_i >$ where U is the mixing matrix and the sum is over the three mass eigenstates states with masses m_1, m_2, and m_3. The time evolution equation of an initial state $|\nu_\alpha >$ over time "t" is given by

$$|\nu_\alpha(t) >= \Sigma_i e^{-iE_i t} U_{\alpha i}|\nu_i > , \tag{2.71}$$

where E_i stands for the energy of the i-th mass eigenstate. We define the mass squared difference Δ_{ij}^2 as $\Delta_{ij}^2 = m_i^2 - m_j^2$. This gives us the probability of finding a flavor ν_β with an original beam of $|\nu_\alpha >$ travelling a distance "L" with energy "E" in a vacuum as

$$P_{\alpha\beta} = \delta_{\alpha\beta} - 4\Sigma U_{\alpha i} U_{\beta i}^* U_{\alpha j}^* U_{\beta j} sin^2 \left(\frac{1.27\Delta_{ij}^2 L}{E} \right) . \tag{2.72}$$

The above equation is correct for neutrinos travelling in a vacuum. However, the neutrinos are created deep inside the SN core at very high density and they come out through a very large density gradient. During that, all three flavors interact with matter through the neutral current whereas the electron type ν_e and $\bar{\nu}_e$ interact with the charged current interaction as well, picking up an additional mass term. This term for the charged current on ν_e inside the SN is given by Choubey and Kar [108]

$$A(r) = 2\sqrt{2}G_F N_{Av} n_e(r) E_\nu , \qquad (2.73)$$

with G_F the Fermi coupling constant, N_{Av} the Avogadro number, $n_e(r)$ the ambient electron density at radius "r" and E_ν the energy of the neutrino beam. This can be written in the units we are using as

$$\frac{A(r)}{eV^2} = 15.14 \times 10^{-8} Y_e(r) \left[\frac{\rho(r)}{g/cm^3}\right]\left[\frac{E_\nu}{MeV}\right]. \qquad (2.74)$$

As we have three neutrino masses, there are two independent Δm^2 values. The global analysis of the solar data and KamLAND reactor antineutrino disappearance experiment gives $\Delta m_{21}^2 \sim 7.39 \times 10^{-5}\,eV^2$ and $\sin^2\theta_{12} \sim 0.310$. $|\Delta m_{32}^2| \sim 2.449 \times 10^{-3}\,eV^2$, $\sin^2\theta_{23} \sim 0.558$ are obtained from the atmospheric neutrino data but the sign of the mass squared difference is not known definitely. Also the current value of $\sin^2\theta_{13} \sim 0.0224$. So one talks of two mass hierarchies: (1) Normal Hierarchy (NH) with $m_3 > m_2$ and (2) Inverted Hierarchy (IH) with $m_2 > m_3$. So in the presence of matter effect, the number of ν_e, $\bar{\nu}_e$, ν_x and $\bar{\nu}_x$ coming out from the supernova, N_{ν_e}, $N_{\bar{\nu}_e}$, N_{ν_x} $N_{\bar{\nu}_x}$ (here "x" stands for "μ" and "τ"), respectively, are given by Choubey and Kar [108]

$$N_{\nu_e} = P_{ee}(E)N_{\nu_e}^0(t) + (1 - P_{ee}(E))N_{\nu_x}^0(t) , \qquad (2.75)$$

$$N_{\bar{\nu}_e} = \bar{P}_{ee}(E)N_{\bar{\nu}_e}^0(t) + (1 - \bar{P}_{ee}(E))N_{\bar{\nu}_x}^0(t) , \qquad (2.76)$$

$$2\,N_{\nu_x} = (1 - P_{ee}(E))N_{\nu_e}^0(t) + (1 + P_{ee}(E))N_{\nu_x}^0(t) , \qquad (2.77)$$

$$2\,N_{\bar{\nu}_x} = (1 - \bar{P}_{ee}(E))N_{\bar{\nu}_e}^0(t) + (1 + \bar{P}_{ee}(E))N_{\bar{\nu}_x}^0(t) , \qquad (2.78)$$

where the probabilities P_{ee} and $\bar{P}_{ee}$ are the survival probabilities of ν_e and $\bar{\nu}_e$ and given by Choubey and Kar [108], and $N_{\nu_\alpha}^0(t) = L_{\nu_\alpha}(t)/ < E_\alpha(t) >$ where $L_{\nu_\alpha}(t)$ and $< E_{\nu_\alpha}(t) >$ are the luminosity and the average energy of the ν_α at time "t." The above equations use the fact that the beam of ν_μ and ν_τ cannot be distinguished from each other and both are denoted by ν_x with $N_{\nu_\mu}^0 = N_{\nu_\tau}^0 = N_{\nu_x}^0$. Similarly $\bar{\nu}_x$ stands for $\bar{\nu}_\mu$ and $\bar{\nu}_\tau$.

The stellar core during the supernova evolution reach such high densities of matter that even the neutrino-neutrino interaction becomes important leading to a complex phenomenon of collective neutrino oscillations. This has effects on the spectra of the neutrinos and leads to spectral swaps or spectral split features, as a result of which flavors are effectively swapped above or below an energy threshold. The effect of collective oscillations on possible future supernova neutrinos are extensively studied in the last 15 years [85, 109, 110]. But here we shall restrict ourselves to discussion on enhanced neutrino oscillations through matter effect and not with collective oscillations on top of it.

The fluxes of the four types of neutrino/antineutrinos ν_e, $\bar{\nu}_e \nu_x$ and $\bar{\nu}_x$ are often given in a parametrized form as

$$f(E_\nu) = N \left(\frac{E_\nu}{\bar{E}_\nu}\right)^\alpha exp^{[-\frac{(\alpha+1)E_\nu}{\bar{E}_\nu}]} , \qquad (2.79)$$

where $\bar{E}_\nu$ stands for the mean energy of the neutrinos of any flavor. The parameter "α" (different from the flavor index) is often called the pinching parameter as it reduces the fluxes at higher energies. The normalization constant N is given by $N = \frac{(\alpha+1)^{\alpha+1}}{\bar{E}_\nu \Gamma(\alpha+1)}$ where $\Gamma(x)$ is the Gamma function. The fluxes along with the total energies, average energies, and the α's are all functions of time, separate ones for the four types of neutrinos/antineutrinos.

The neutrinos get emitted in three distinct phases as described below. In the neutronization phase lasting for about 20–30 ms, initially ν_e emission is dominant but then all flavors of $\nu/\bar{\nu}$ come out. Then in the accretion phase lasting for about 200–700 ms, all flavors of $\nu/\bar{\nu}$ are emitted with the average energies $\bar{E}_{\nu_x} > \bar{E}_{\bar{\nu}_e} > \bar{E}_{\nu_e}$. Finally, in the cooling phase lasting for about 10 s, the average energies all go down with time, and eventually the average energies of all flavors roughly become equal.

The $\nu/\bar{\nu}$ interact with the matter in the detector in four different ways:

1. The inverse beta decay (IBD) process on protons where the electron type antineutrinos interact with the protons as $\bar{\nu}_e + p \rightarrow n + e^+$, with the kinematic threshold of 1.8 MeV, is the first option. The positron's energy loss mostly can be observed and one sees approximately $E_{e^+} = E_{\bar{\nu}} - 1.3$ MeV. In scintillators, the 0.511-MeV positron annihilation gamma rays can also be seen. The neutron can be captured by free protons producing a deuteron and a 2.2 MeV gamma. It can also get absorbed by another nucleus. If the detector is doped with a substance with a large neutron capture cross section like gadolinium (Gd) in the water, then the detection becomes more effective.
2. The elastic scattering on electrons in the detector given by $\nu_\alpha + e^- \rightarrow \nu_\alpha + e^-$ for all neutrino/antineutrino flavors "α" is an important process as the cross section is forward peaked in the direction of the incoming neutrino. It can be used for pointing the direction of the supernova event. The ν_e and $\bar{\nu}_e$ interact through both CC and NC process whereas ν_x interact through NC only. However, the cross sections are smaller compared to the IBD.
3. The CC and the NC interactions on nuclei is the third possibility. The $\nu_e/\bar{\nu}_e$ interact with nuclei via the reaction $\nu_e + (N, Z) \rightarrow (N - 1, Z + 1) + e^-$ and $\bar{\nu}_e + (N, Z) \rightarrow (N + 1, Z - 1) + e^+$. The kinematic threshold of both the

reactions is $E_\nu^{th} = (M_f^2 + m_e^2 + 2M_f M_e - M_i^2)/2M_i \sim (M_f - M_i + m_e)$, where M_f and M_i are the final and initial masses of the nuclei involved. The ν_μ and ν_τ and their antiparticles do not have CC interactions as their energies are much smaller than the masses of μ and the τ particles. In this case the loss of energy of charged leptons is observable. Also the ejecta (nucleons and gammas) produced by the final nuclei during their de-excitation also can be observed.

4. The coherent elastic neutrino-nucleus scattering of the detector nuclei and protons through NC interactions is an important detection process. The interaction rate is comparatively high but recoil energies are low and so one needs to have very low threshold detectors. The differential number of neutrino events at the detector for a given reaction process is given by

$$\frac{d^2 S}{dE_\nu dt} = \Sigma_\alpha \frac{n}{4\pi D^2} N_{\nu_\alpha} f_{\nu_\alpha}(E_\nu) \sigma_{\nu_\alpha}(E_\nu) \epsilon_{\nu_\alpha}(E_\nu) , \qquad (2.80)$$

where α runs over the neutrino species, N_{ν_α} is the neutrino flux as given by Eqs. (2.75)–(2.80). D is the distance of the source from the detector as earlier, $\sigma_\alpha(E_\nu)$ is the reaction cross section with a target particle, n is the number of target particles present, $f_\alpha(E_\nu)$ stands for the energy spectrum of the neutrinos of flavor "α" and $\epsilon_\alpha(E_\nu)$ is the efficiency of the detector depending on the neutrino energy. This when integrated over the neutrino energy and time gives us the total number of events for the detector.

Finally, we describe the detectors already in existence or are planned for the supernova neutrinos. As one wants to have as many events as possible, the strategy is to have detectors of large mass. The detectors can be classified as:

(A) Scintillation Detectors
Scintillation detectors are made of hydrocarbons, with the approximate chemical formula $C_n H_{2n}$. The energy loss is through the de-excitation of molecular energy levels. The detectors often are large volumes of liquid observed by photomultiplier tubes. The IBD on the protons of the hydrocarbons is essentially the mode for the neutrino detection reaction. As is shown in Table 2.2, many detectors like KamLAND in Japan, Borexino in Italy, LVD in Italy, SNO+ in SNOLAB, Canada all use scintillators. Electron scattering gives only a few percent of the events. The neutrinos can excite the ^{12}C too- NC excitation of ^{12}C followed by the emission of a 15.1 MeV gamma ray can be seen and with a good energy resolution of the detectors, one can estimate the flux of the neutrinos.

(B) Water Cerenkov Detectors
Water is one of the most inexpensive materials for making large neutrino detectors as it has protons available for the IBD process. Charged particles are detected by their Cherenkov light emission observed by the PMTs and which forms a 42^0 cone for relativistic particles. Dissolving Gd compounds in the water enhance neutron tagging possibility. Earlier detectors like K II in Japan and IMB in the USA were water Cherenkov detectors as is the present SK in Japan. The nucleus ^{16}O can also

Table 2.2 Expected total neutrino events (rounded off) in some prominent present and future supernova neutrino detectors for a typical galactic SN at a distance of 10 kpc. The number of events are taken from [106] with the "GKVM"[116] model for neutrino emission except for DUNE which is from [115]

Detector	Year started	Type	Mass (kt)	Total events
LVD	1992	$C_n H_{2n}$	1	300
Super-Kamiokande	1996	$H_2 O$	32	7000
KamLAND	2002	$C_n H_{2n}$	1	300
MiniBooNE	2002	$C_n H_{2n}$	0.7	200
Borexino	2005	$C_n H_{2n}$	0.3	100
HALO	2012	Pb	0.08	30
SNO+	2014	$C_n H_{2n}$	0.8	300
NovA	2014	$C_n H_{2n}$	15	4000
Icarus	Near future	Ar	0.6	60
DUNE	Future	Ar	40	3700
Hyper-Kamiokande	Future	$H_2 O$	540	110,000
LENA	Future	$C_n H_{2n}$	50	15,000

be excited by the neutrinos with a de-excitation through gamma rays. The Sudbury Neutrino Observatory (SNO) used heavy water and through the CC and NC breakup reactions could detect neutrinos of all flavors. It had 1 kton of heavy water along with 1.7 kton of light water.

(C) Liquid Argon Detectors
Liquid Argon Time Projection Chambers (TPC) is important as they can observe supernova neutrinos through $\nu_e +^{40} Ar \rightarrow e^- +^{40} K^*$. NC interactions are also possible though their reaction cross section is not known that accurately. In principle, the de-excitation of $^{40}K^*$ through the γs can be tagged. The electron type antineutrino capture reaction $\bar{\nu}_e +^{40} Ar \rightarrow e^+ +^{40} Cl^*$ and the elastic scattering of neutrinos on electrons can also be observed. A present example of this type of detector is ICARUS in Italy and a future prospect is the DUNE with its near and far detection facilities in the USA.

(D) Heavy Nucleus Detectors
If the detector is made of a heavy nucleus like lead, then the CC interaction reaction like $\nu_e +^{208} Pb \rightarrow e^- +^{208} Bi$, or the NC interaction will excite the final nucleus beyond particle emission threshold and then it will de-excite through the emission of neutrons, protons or gammas. One plan is to detect the emitted neutrons and from the measurement of the neutron energies, reconstruct the neutrino spectrum. ^{56}Fe is another possibility but it emits far less neutrons than ^{208}Pb [111]. The Detector HALO uses lead as detector material and is housed in the SNOLAB in Canada. Liquid xenon detectors, constructed for dark matter detections can also detect supernova neutrinos, in principle [112].

Table 2.2 gives the total number of events for a core collapse supernova at a distance of 10 kpc that can be observed in some of the present and future detectors across the globe. One should note that these numbers depend strongly on the model of neutrino production and emission from the source. For example, for DUNE, the number of events for the "Livermore" model [113] and the "Garching" model [114] models are 3213 and 1047, respectively, [115] with no flavor mixing whereas the number 3656 given in Table 2.2 is for the "GKVM" model [116] including collective effects. But the important thing that this table shows is how today the number of supernova neutrinos that can be observed has increased by more than an order of magnitude compared to the year 1987 taking into account the fact that the numbers in the table are for a distance of 10 kpc whereas the distance of LMC for SN1987A is about 50 kpc; also once Hyper-Kamiokande starts operating, this increase will be much more.

If one is able to determine the direction of the future supernova event from the neutrino observation, it will be of paramount importance. The neutrino burst is expected to hit the earth a few hours before the electromagnetic radiation reaches and so all the telescopes/observational facilities can look for that part of the sky. The directionality is observed mostly through elastic scattering of electrons which is a few percent of the signal. After taking background into account, Super Kamiokande expects its pointing quality at about 8^0 for 10 kpc distance, which can be improved to about 3^0 with good inverse beta decay tagging. It should be mentioned here that today the SuperNova Early Warning System (SNEWS) is ready which is an international network of detectors that plans to provide an alert to astronomers.

Finally, we mention an exciting idea connected to the emission of supernova neutrinos, known as Diffuse Supernova Neutrino Background (DSNB). It is the total population of neutrinos and antineutrinos originating from all supernova in the cosmic history of the universe [117–119]. Theoretical quantitative prediction of DSNB has many uncertainties coming from the star formation rate, the rate of supernova events in the universe as a function of time, the neutrino spectra from each supernova, the oscillation of neutrino flavors, etc. [119] Still one is able to calculate the expected rate; for the $\bar{\nu}_e$ component of the DSNB the estimated value is 0.1 to 1.0 cm^{-2} s^{-1} for energies above about 19 MeV [106]. Current experimental limit from the Super Kamiokande (SK) experiment is seen to be 2.8–3.0 $\bar{\nu}_e$ cm^{-2} s^{-1} for antineutrino energies above 17.3 MeV [118]. The background of solar and reactor neutrino events below energies $\sim$17 MeV and atmospheric neutrino events above energies $\sim$40 MeV give only a window, of roughly 20 MeV for observing the DSNB for SK. But still with all these uncertainties, the DSNB is close to the possibility of experimental detection in the near future.

To conclude, we stress that the neutrino detectors globally are ready to observe the next galactic supernova and learn exciting things about the supernova mechanisms and properties of neutrinos.

References

1. Bethe, H.A.: Supernova mechanisms. Rev. Mod. Phys. **62**, 801 (1990)
2. Arnett, D., Bahcall, J.N., Krishner, R.P., Woosley, S.E.: Supernova 1987A. Ann. Rev. Astron. Astrophys. **27**, 629–700 (1989)
3. Woosley, S.E., Heger, A: The death of very massive stars. In: Vink, J.S. (ed.) Very Massive Stars in the Local Universe, pp. 199–225. Springer, Berlin (2019)
4. Arnould, M., Takahashi, K.: Nuclear astrophysics. Rep. Prog. Phys. **62**, 395 (1999)
5. Janka, H.-T., Hanke, F., Hudelpohl, L., Marek, A., Müller, B., Obergaulinger, M.: Core-collapse supernovae: reflections and directions. Prog. Th. Expt. Phys. **2012**(1), 01A309 (2012)
6. Saha, M.N.: Ionisation in solar chromosphere. Phil. Mag. Sr. VI **40**, 472 (1920)
7. Saha, M.N.: On a physical theory of stellar spectra. Proc. Roy. Soc. Lond. **A99**, 135 (1921)
8. Clayton, D.D.: Principles of Stellar Evolution and Nucleosynthesis. The University of Chicago Press, Chicago (1983)
9. Iliadis, C.: Nuclear Physics of Stars. Wiley, Hoboken (2007)
10. Chandrasekhar, S.: The maximum mass of ideal white dwarfs. Astrophys. J. **74**, 81 (1931)
11. Chandrasekhar, S.: An Introduction to the Study of Stellar Structure. The Chicago University Press, Chicago (1939)
12. Bethe, H.A., Brown, G.E., Applegate, J., Lattimer, J.M.: Equation of state in the gravitational collapse of stars. Nucl. Phys. A **324**, 487–533 (1979)
13. Kar, K., Ray, A., Sarkar, S.: Beta decay rates of FP shell nuclei with $A > 60$ in massive stars at the presupernova stage. Astrophys. J. **434**, 662 (1994)
14. Goldreich, P., Weber, S.V.: Homologously collapsing stellar cores. Astrophys. J. **238**(1), 991–997 (1980)
15. Yahil, A., Lattimer, J.M.: Supernovae for pedestrians. In: Rees, M.J., Stoneham R.J. (eds.) Supernovae, pp 57–70. Reidel, Dordrecht
16. Yahil, A.: Self-similar stellar collapse. Astrophys. J. **265**, 1047–1055 (1983)
17. Arnett, D.: Supernova and Nucleosynthesis. Princeton University Press, Princeton (1996)
18. deShalit, A., Feshbach, H.: Theoretical Nuclear Physics- Volume 1. Wiley, New York (1974)
19. Fuller, G.M., Fowler, W.A., Newman, M.J.: Stellar weak interaction rates for sd-shell nuclei. I. Nuclear matrix element systematics with application to ^{26}Al and selected nuclei of importance to the supernova problem. Astrophys. J. Suppl. **42**, 447–473 (1980)
20. Fuller, G.M., Fowler, W.A., Newman, M.J.: Stellar weak interaction rates for intermediate-mass nuclei. II- A=21 to A=60. Astrophys. J. **252**, 715–740 (1982)
21. Fuller, G.M., Fowler, W.A., Newman, M.J.: Stellar weak reaction rates for intermediate-mass nuclei. IV. Interpolation procedures for rapidly varying lepton capture rates using effective log (ft)-values. Astrophys. J. **293**, 1–16 (1985)
22. Caurier, E., Langanke, K., Martinez-Pinedo, G., Nowacki, F.: Shell-model calculations of stellar weak interaction rates. I. Gamow-Teller distributions and spectra of nuclei in the mass range A=45-65. Nucl. Phys. A **653**(4), 439–452 (1999)
23. Martinez-Pinedo, G., Langanke, K., Dean, D.J.: Competition of electron capture and beta-decay rates in supernova collapse. Astrophys. J. Suppl. **126**, 493–499 (2000)
24. Koonin, S.E., Dean, D.J., Langanke, K.: Shell model Monte Carlo methods. Phys. Rep. **278**(1), 1–77 (1997)
25. Bruenn, S.W.: Stellar core collapse- numerical model and infall epoch. Astrophys. J. Suppl. **58**, 771–841 (1985)
26. Rampp, M., Janka, H.-T.: Radiation hydrodynamics with neutrinos- Variable Eddington factor method for core collapse supernova simulations. Astron. Astrophys. **398**, 361 (2002)
27. Langanke, K., Martinez-Pinedo, G.: Nuclear weak-interaction processes in stars. Rev. Mod. Phys. **75**(3), 819 (2003)
28. Janka, H.-T., Langanke, K., Marek, A., Martinez-Pinedo, G., Müller, B.: Theory of core-collapse supernovae. Phys. Rep. **442**, 38–74 (2007)
29. Freedman, D.Z.: Coherent effects of a weak neutral current. Phys. Rev. D **9**(5), 1389 (1974)

30. Mazurek, T.: Chemical potential effects on neutrino diffusion in supernovae. Astrophys. Space Sci. **35**, 117–135 (1975)
31. Sato, K.: Supernova explosion and neutral currents of weak interaction. Prog. Theor. Phys. **54**(5), 1325–1338 (1975)
32. Tubbs, D.L., Schramm, D.N.: Neutrino opacities at high temperatures and densities. Astrophys. J. **201**, 467–488 (1975)
33. Lamb, D.Q., Pethick, C.J.: Effects of neutrino degeneracy in supernova models. Astrophys. J. **209**, L77–L81 (1976)
34. Cooperstein, J., Wambach, K.: Electron capture in stellar collapse. Nucl. Phys. A **420**(3), 591–620 (1984)
35. Lattimer, J.L., Pethick, C.J., Ravenhall, D.J., Lamb, D.Q.: Physical properties of hot, dense matter: the general case. Nucl. Phys. A **432**(3), 646–742 (1985)
36. Ravenhall, D.J., Bennett, C.D., Wilson, J.R.: Structure of matter below nuclear saturation density. Phys. Rev. Lett. **50**, 646 (1983)
37. Cooperstein, J.: The equation of state in supernovae. Nucl. Phys. A **438**, 722–739 (1985)
38. Baron, E., Cooperstein, J., Kahana, S.: Type II supernovae in 12 $M_\odot$ and 15 $M_\odot$ stars: the equation of state and general relativity. Phys. Rev.Lett. **55**(1), 126 (1985)
39. Baron, E., Cooperstein, J., Kahana, S.: Supernovae and the nuclear equation of state at high densities. Nucl. Phys. A **440**(4), 744–754 (1985)
40. Blaizot, J.P., Gogny, D., Grammaticos B.: Nuclear compressibility and monopole resonances. Nucl. Phys. A **265**(2), 315–336 (1976)
41. Brown, G.E.: Dense nuclear matter: supernovae and heavy ions. Zeit. Phys. C **38**, 3 (1988)
42. Pines, D., Quader, K.F., Wambach, J.: Effective interactions and elementary excitations in nuclear matter. Nucl. Phys. A **477**(3), 365–398 (1988)
43. Stone, J.R., Stone, N.J., Moszkowski, S.: Incompressibility in finite nuclei and nuclear matter. Phys. Rev. C **89**(4), 044316 (2014)
44. Piekarewicz, J.: Unmasking the nuclear matter equation of state. Phys. Rev. C **69**(4), 041301 (2004)
45. Shlomo, S., Kolomietz, V. M., Coló, G.: Symmetry energy from the nuclear collective motion. Eur. Phys. J. A **30**, 23 (2006)
46. Bethe, H.A., Wilson, J.R.: Revival of stalled supernova shock by neutrino heating. Astrophys. J. **295**, 14–23 (1985)
47. Bowers, R.L., Wilson J.R.: A numerical model for stellar core collapse calculations. Astrophys. J. Suppl. **50**, 115–159 (1982)
48. Murphy, J.W., Burrows, A.: Criteria for core-collapse supernova explosions by the neutrino mechanism. Astrophys. J. **688**, 1159–1175 (2008)
49. Burrows, A., Vartanyan, D., Dolence, J.C., Skinner, M.A. Radice, D.: Crucial physical dependencies of the core-collapse supernova mechanism. Space Sci. Rev. **214**, 33 (2018)
50. O'Connor, E., Ott, C.D.: Black hole formation in failing core-collapse supernovae. Astrophys. J. **730**, 70 (2011)
51. O'Connor, E., Ott, C.D.: A new open source code for spherically symmetric stellar collapse to neutron stars and black holes. Class. Q. Grav. **27**, 114103 (2010)
52. Gourgoulhon, E.: Simple equations for general relativistic hydrodynamics in spherical symmetry applied to neutron star collapse. Astron. Astrophys. **252**, 651–665 (1991)
53. Noble, S.C.: A numerical study of relativistic fluid collapse. Ph.D. Thesis, University of British Columbia (2003)
54. Banyuls, F., Font, J.A., Ibanez, J.M., Marti, J.M., Miralles, J.A.: Numerical (3+1) general relativistic hydrodynamics: a local characteristic approach. Astrophys. J. **476**, 221–231 (1997)
55. Romero, J.V., Ibanez, J.M., Marti, J.M., Miralles, J.A.: A new spherically symmetric general relativistic hydrodynamical code. Astrophys. J. **462**, 839 (1996)
56. Liebendörfer, M.: A simple parametrization of the consequences of deleptonization for simulations of stellar core collapse. Astrophys. J. **633**, 1042–1051 (2005)

57. Ruffert, M., Janka, H.-T., Schafer, G.: Coalescing neutron stars- a step towards physical models I. Hydrodynamic evolution and gravitational-wave emission. Astron. Astrophys. **311**, 532–566 (1996)
58. Rosswog, S., Liebendörfer, M.: High-resolution calculations of merging neutron stars II. Neutrino emission. Mon. Not. R. Astron. Soc. **342**(3), 673–689 (2003)
59. Janka, H.-T.: Conditions for shock revival by neutrino heating in core-collapse supernovac. Astron. Astrophys. **368**, 527 (2001)
60. Fischer, T., Whitehouse, S.C., Mezzacappa, A., Thielemann, F.-K., Liebendörfer, M.: The neutrino signal from protoneutron star accretion and black hole formation. Astrophys. J. **499**, 1 (2009)
61. Sumiyoshi, K., Yamada, S., Suzuki, H.: Dynamics and neutrino signal of black hole formation in nonrotating failed supernovae I. Equation of state dependence. Astrophys. J. **667**, 382–394 (2007)
62. Marek, A., Janka, H.-T.: Delayed neutrino-driven supernova explosions aided by the standing accretion-shock instability. Astrophys. J. **694**, 664–696 (2009)
63. Afle, C., Brown, D.A.: Inferring physical properties of stellar collapse by third-generation gravitational-wave detectors. Phys. Rev. D **103**, 023005 (2021)
64. Schneider, A.S., Roberts, L.F., Ott, C.D., O'Connor, E.: Equation of state effects in the core collapse of a 20 $M_\odot$ star Phys. Rev. C **100**(5), 055802 (2019)
65. Nagakura, H., Iwakami, W., Furusawa, S., et al.: Simulation of core-collapse supernovae in spatial axisymmetry with full Boltzmann neutrino transport. Astrophys. J. **854**, 136 (2018)
66. Morozovai, V., Radice, D., Burrows, A., Vartanyan, D.: The gravitational wave signal from core-collapse supernovae. Astrophys. J. **861**, 10 (2018)
67. Richers, S., Ott, C.D., Abdikamalov, E., O'Connor, E., Sullivan, E.: Equation of state effects on gravitational waves from rotating core collapse. Phys. Rev. D **95**(6), 063019 (2017)
68. Furusawa, S., Sumiyoshi, K., Yamada, S., Suzuki, H.: Supernova equations of state including full nuclear ensemble in-medium effects. Nucl. Phys. A **957**, 188–207 (2017)
69. Char, P., Banik, S., Bandyopadhyay, D.: A comparative study of hyperon equation of state in supernova simulations. Astrophys. J. **809**, 116 (2015)
70. Fischer, T., Hempel, M., Sagert, I., Suwa, Y., Schaffner-Bielich, J.: Symmetry energy impact in simulations of core collapse supernovae. Eur. Phys. J. A **50**, 46 (2014)
71. Couch, S.M.: The dependence of the neutrino mechanism of core-collapse supernovae on equation of state. Astrophys. J. **765**, 29 (2013)
72. Hempel, M., Fischer, T., Schaffner-Bielich, J., Liebendörfer, M.: New equations of state in simulations of core-collapse supernovae. Astrophys. J. **748**, 70 (2012)
73. Sagert, I., Fischer T., Hempel M., et al.: Signals of the QCD phase transition in core-collapse supernovae. Phys. Rev. Lett. **102**, 081101 (2009)
74. Banik, S., Hempel, M., Bandyopadhyay, D.: New hyperon equations of state for supernovae and neutron stars in density-dependent hadronic field theory. Astrophys. J. **214**, 22 (2014)
75. Shen, H., Toki, H., Oyamatsu, K., Sumiyoshi, K.: Relativistic equation of state for core-collapse supernova simulations. Astrophys. J. Suppl. **197**, 20 (2011)
76. Brown, G.E., Bethe, H.A.: A scenario for a large number of low-mass black holes in the galaxy. Astrophys. J. **423**, 659 (1994)
77. Keil, W., Janka, H.-T.: Hadronic phase transition at supranuclear densities and the delayed collapse of newly formed neutron stars. Astron. Astrophys. **296**, 145 (1995)
78. Baumgarte, T.W., Janka, H.-T., Keil, W., Shapiro, S.L., Teukolsky, S.A.: Delayed collapse of hot neutron stars to black holes via hadronic phase transitions. Astrophys. J. **468**, 823–833 (1996)
79. Banik, S.: Role of hyperons in black hole formation. J. Phys. Conf. Ser. **426**, 012004 (2013)
80. Mayle, R., Wilson, J.R.: Supernovae from collapse of oxygen-magnesium-neon cores. Astrophys. J. **334**, 909–926 (1988)
81. Kitaura, F.S., Janka, H.-T., Hillebrandt, W.: Explosions of O-Ne-Mg cores, the Crab supernova and subluminous type II-P supernovae. Astron. Astrophys. **450**, 345–350 (2006)

82. Burrows, A., Sawyer, R.F.: Effects of correlations on neutrino opacities in nuclear matter. Phys. Rev. C **58**(1), 554 (1998)
83. Epstein, A.T.: Lepton-driven convection in supernovae. Mon. Not. R. Astron. Soc. **188**(2), 305-325 (1979)
84. Burrows, A., Lattimer J.M.: Neutrinos from SN 1987A. Astrophys. J. Lett. **318**, L63–68 (1987)
85. Horiuchi, S., Kneller, J.P.: What can be learned from a future supernova neutrino detection? J. Phys. G Nucl. Part. Phys. **45**, 043002 (2018)
86. Blondin, J.M., Mezzacappa, A., De Marino, C.: Stability of standing accretion shocks, with an eye toward core collapse supernovae. Astrophys. J. **584**, 971 (2003)
87. Janka, H,-T., Langanke, K., Marek, A., Martinez-Pinedo, G., Müller, B.: Theory of core-collapse supernovae. Phys. Rep. **492**(1–6), 38–74 (2007)
88. Hannestad, S., Raffelt, G.: Supernova neutrino opacity from nucleon-nucleon Bremsstrahlung and related processes. Astrophys. J. **507**, 339–352 (1998)
89. Burrows, A., Vartanyan, D.: Core-collapse supernova explosion theory. Nature **589**, 29–31 (2021)
90. Couch, S.M., Ott, C.D.: The role of turbulence in neutrino-driven core-collapse supernova explosions. Astrophys. J. **799**, 5 (2015)
91. Skinner, M.A., Dolence, J.C., Burrows, A., Radice, D., Vartanyan, D.: Fornax: a flexible code for multiphysics astrophysical simulations. Astrophys. J. Suppl. **241**, 7 (2019)
92. Nagakura, H., Burrows, A., Radice, D., Vartanyan, D.: Towards an understanding of the resolution dependence of core-collapse supernova simulations. Mon. Not. R. Astron. Soc. **490**(4), 4622–4637 (2019)
93. Janka, H.-T., Melson, T., Summa, A.: Physics of of core-collapse supernovae in three dimensions; a sneak preview. Ann. Rev. Nucl. Part. Sci. **66**, 341 (2016)
94. Müller, B.: Hydrodynamics of core-collapse supernovae and their progenitors. Living Rev. Comput. Astrophys. **6**, 3 (2020)
95. Dessart, L., Hillier, J.J., Wilk, K.D.: Impact of clumping on core-collapse supernovae. Astron. Astrophys. **619**, A30 (2018)
96. Cigan, P., Matasura, M., Gomez, H.L., et al.: High angular resolution ALMA images of dust and molecules in the SN 1987A ejecta. Astrophys. J. **886**, 51 (2019)
97. Page, D., Beznogov, M.V., Garibay, I., Lattimer, J.M., Prakash, M., Janka, H.-T.: NS 1987A in SN 1987A. Astrophys. J. **898**, 125 (2020)
98. Bionta, R.M., Blewitt, G., Bratton, C.B., et al.: Observation of a neutrino burst in coincidence with supernova 1987A in the large magellanic cloud. Phys. Rev. Lett. **58**, 1494 (1987)
99. Hirata, K., Kajita, T., Nakahata M., et al.: Observation of a neutrino burst from the supernova SN1987A. Phys. Rev. Lett. **58**, 1490 (1987)
100. Alexeyev, E.N., Alexeyeva, L.N., Krivosheina, I.V., Volchenko, V.I.: Detection of the neutrino signal from SN1987A in the LMC using the INR Baksan underground scintillation telescope. Phys. Lett. B **205**(2–3), 209–214 (1988)
101. Aglietta, M., Badino, G., Bologna, G., et al.: Comments on the two events observed in neutrino detectors during the supernova 1987a outburst. Europhys. Lett. **3**, 1321 (1987)
102. Kolb, E.W., Stebbins, A.J., Turner, M.S.: How reliable are neutrino mass limits derived from SN1987A? Phys. Rev. D **35**, 3598–3606 (1987)
103. Mohapatra, R.N., Pal, P.B.: Massive Neutrinos in Physics and Astrophysics. World Scientific, Singapore (2004)
104. Burrows, A., Lattimer, J.M.: Neutrinos from SN 1987A. Astrophys. J. **318**, L63–L68 (1987)
105. Kar, K.: Neutrinos from SN1987A- observations and expectations. In: Ray, A., Velusamy T. (eds.) Proceedings of School and Workshop on Supernovae and Stellar Evolution, pp. 222–233. World Scientific, Singapore (1991)
106. Scholberg, K.: Supernova neutrino detection. Ann. Rev. Nucl. Part. Sc. **62**, 81–103 (2012)
107. Mirrizi, A., Raffelt, G.G., Serpico, P.D.: Earth matter effects in supernova neutrinos: optimal detector locations. J. Cosmol. Astropart. Phys. **0605**, 012 (2006)
108. Choubey, S., Kar, K.: Neutrinos from supernovae. Proc. Ind. Nat. Sci. Acad. **70**, A 123 (2004)

109. Duan, H., Fuller, G.M., Qian, Y.-Z.: Collective neutrino oscillations. Ann. Rev. Nucl. Part. Sci. **60**, 569–594 (2010)
110. Chakraborty, S., Hanseni, R., Izaguirre, I., Raffelt, G.G.: Collective neutrino flavor conversion: recent developments. Nucl. Phys. B **908**, 366–381 (2016)
111. Bandyopadhyay, A., Bhattacharjee, P., Chakraborty, S., Kar, K., Saha, S.: Detecting supernova neutrinos with iron and lead detectors. Phys. Rev. D **95**, 065002 (2017)
112. Bhattacharjee, P., Bandyopadhyay, A., Chakraborty, S., Ghosh, S., Kar, K., Saha, S.: Supernova neutrino induced neutrons in liquid xenon dark matter detectors. arxiv:2012.14888 (2020)
113. Totani, T., Sato, K., Dalhed, H.E., Wilson, J.R.: Future detection of supernova neutrino burst and explosion mechanism. Astrophys. J. **496**, 216–225 (1998)
114. Hüdelpohl, L., Müller, B., Janka, H.-T., Marek, A., Raffelt, G.G.: Neutrino signal of electron-capture supernovae from core collapse to cooling. Phys. Rev. Lett. **104**, 251101 (2010)
115. Abi, B., Acciarri, R., Acero, M.A., et al.: Supernova neutrino burst detection with the deep underground neutrino experiment. arXiv:2008.06647 (2020)
116. Gava, G., Kneller, J., Volpe, C., McLaughlin, G.C.: Dynamical collective calculation of supernova neutrino signals. Phys. Rev. Lett. **103**, 071101 (2009)
117. Beacom, J.F.: The diffuse supernova neutrino background. Ann. Rev. Nucl. Part. Sci. **60**, 439–462 (2010)
118. Zhang, H., Abe, K., Hayato, Y., et al.: Supernova relic neutrino search with neutron tagging at Superkamiokande-IV. Astropart. Phys. **60**, 41–46 (2015)
119. Mirrizi, E., Tamborra, I., Janka, H.-T., et al.: Supernova neutrinos: production, oscillations and detection. Riv. Nuovo Cim. **39**(1–2), 1 (2016)

Chapter 3
Neutron Stars

Summary Neutron stars are unique laboratories under astrophysical conditions of extreme matter densities, strongly quantizing magnetic fields, and strong gravity. These compact objects encompass the density of a Fe nucleus in the crust and several times the normal nuclear matter density (2.7×10^{14} g/cm^3) in the core. Many interesting phases of matter are hypothesized to exist in neutron stars. On the other hand, very strong surface magnetic fields $\sim 10^{16}$ G are found in a class of neutron stars known as magnetars. Observations of neutron stars across the electromagnetic spectrum lead to the estimation of their masses, radii, and moments of inertia. Observed gross properties of neutron stars constrain the compositions and EoS of dense matter modelled theoretically.

3.1 History and Discovery of Neutron Stars

The life cycle of a star would either evolve to a white dwarf or a neutron star/black hole. The outcomes of main sequence stars depend on the masses of progenitor stars. This is discussed in the preceding chapter on the theory of supernova explosions. We denote white dwarfs and neutron stars as compact stars. Sirius B was the most studied white dwarf since the second half of the nineteenth century. With the advent of the Fermi-Dirac statistics, the study of cold and compressed matter in white dwarfs attracted the attention of theoretical physicists [1, 2]. R. H. Fowler argued that the pressure of degenerate electrons was responsible for the stability of a white dwarf against the gravitational collapse [3]. Later Subrahmanyan Chandrasekhar carried forward Fowler's calculation to the situation of relativistic degenerate electrons and predicted the limiting mass of white dwarfs [4, 5]. Can there be a stable branch of compact stars beyond the stable family of white dwarfs? Any answer to this question was impossible before the discovery of the neutron by J. Chadwick [6].

© The Author(s), under exclusive license to Springer Nature Switzerland AG 2022
D. Bandyopadhyay, K. Kar, *Supernovae, Neutron Star Physics and Nucleosynthesis*, Astronomy and Astrophysics Library,
https://doi.org/10.1007/978-3-030-95171-9_3

However, Lev Landau predicted another kind of compact stars as "giant nucleus" in 1931 in anticipation of the fate of massive stars and it was published [7] two days after the publication of the neutron discovery paper on 27 February 1932. Though this was accidental, Landau's idea might not have been associated with neutron stars at that time [1]. However, the potential of Landau's idea in the context of neutron stars was realized after the discovery of the neutron [2].

The first prediction of neutron stars immediately followed the discovery of the neutron. W. Baade and F. Zwicky said in the Stanford meeting of the American Physical Society (APS) in 1933 that the transition from ordinary stars to neutron stars happened in supernova explosions [8]. This was an important milestone in the research of neutron stars. Further J. Chadwick's discovery of the neutron initiated the research on the role of neutrons in the stellar matter. S. Flügge dealt with the neutronization due to the capture of an electron by a proton and a star only made of neutrons [9]. This was followed by the work of F. Hund which discussed how electrons and nuclei together would form neutrons at high pressure and applied his idea to the "dense star" or white dwarf [10]. D.S. Kothari also investigated the neutronization process through the capture of electrons by protons or nuclei in the compressed matter in the context of white dwarfs [11]. In 1939, R. C. Tolman as well as J. R. Oppenheimer and his student G. M. Volkoff published two milestone research articles in the Physical Review that involved the derivation of the hydrostatic equation of a spherically symmetric star using the general relativity and treating neutrons as a Fermi gas. This is now well known as the Tolman–Oppenheimer–Volkoff (TOV) equation for the study of neutron stars [12, 13].

Though all those theoretical activities on cold and dense matter in compact stars happened in the 1930s, it took another 30 years for the observational evidence of neutron stars. The prelude to this journey began with the joining of Antony Hewish in Martin Ryle's research group at the University of Cambridge in 1948 [14]. His main focus was the scintillation phenomenon due to the propagation of radiation originating from radio sources through the plasma medium. He made a detailed plan of a radio telescope for a larger survey of radio galaxies for this purpose in 1965. Jocelyn Bell, a graduate student, joined this program in the same year. The telescope started its operation for the survey as soon as it was ready in 1967. Since August 1967, Bell and Hewish observed something unusual about the fluctuating signals from a faint source. However, it was confirmed on 28 November 1967 that the source was emitting radio pulses at regular intervals. This was the serendipitous discovery of the first pulsar known as CP1919 with a period of 1.33 s. This discovery was reported along with the detections of another three pulsars in the journal in 1968 [15]. Later T. Gold explained that this hitherto unknown astrophysical object was nothing but a rotating neutron star [16].

3.2 Observational Constraints on Neutron Stars

The study of neutron stars gained momentum with the discoveries of the first pulsars by A. Hewish and J. S. Bell. Presently there are about 3000 pulsars known to us. These astrophysical compact objects are now being observed in different wavelengths of the electromagnetic spectrum as well as in gravitational waves. The rapid accumulation of data in recent times provided crucial information about masses, radii, and moments of inertia of neutron stars[17–19]. This, in turn, reveals fundamental aspects of cold and dense matter beyond the normal nuclear matter density. We discuss the progress made in the determination of those observable quantities in the following subsections.

3.2.1 Mass

Measurements of masses of neutron stars are possible in binary systems using Kepler's third law. If one binary component is a pulsar, the rotation of the pulsar can be tracked with the help of a technique known as pulsar timing [20]. A quantity called *mass function* can be defined in terms of two Keplerian parameters out of the five precisely determined Keplerian parameters such as orbital period (P), projection of the semi-major axis of the pulsar (($a_p \sin i$)$/c$), eccentricity (e), $T_0 = GM_\odot/c^3$, and longitude of periastron (ω_0),

$$f_p = \frac{2\pi}{P} \frac{(a_p \sin i)}{G} = \frac{(M_c \sin i)^3}{(M_p + M_c)^2}, \tag{3.1}$$

where "i" is the inclination angle, M_p and M_c are masses of the pulsar and the companion, respectively. This equation is not enough to determine two masses independently even if the inclination angle is known. However, the mass function of the other binary component might be available in certain cases, for example, for an optical companion. In this case, the mass of the pulsar is obtained as $M_p = f_p \alpha (1 + \alpha)^2 / \sin^3 i$ with $\alpha = M_p/M_c$. The pulsar mass, clearly, depends on the inclination angle.

The mass determination of a pulsar is extremely accurate in case of highly relativistic binaries [18]. For such binaries, post-Keplerian parameters such as the relativistic periastron advance ($\dot{\omega}$), the orbital decay ($\dot{P}$), gravitational red-shift and time dilation(γ) and range (r) and shape (s) of the Shapiro delay defined below,

$$\dot{\omega} = 3 \left(\frac{P_b}{2\pi} \right)^{-5/3} (T_\odot M)^{2/3} (1 - e^2)^{-1},$$

$$\gamma = e \left(\frac{P_b}{2\pi} \right)^{1/3} (T_\odot)^{1/3} M^{-4/3} M_c (M_p + 2M_c),$$

$$\dot{P}_b = -\frac{192\pi}{5}\left(\frac{P_b}{2\pi}\right)^{-5/3}\left(\frac{73}{24}e^2 + \frac{37}{96}e^4\right)T_\odot^{5/3}M_pM_cM^{-1/3}(1-e^2)^{-7/2},$$

$$r = T_\odot M_c,$$

$$s = a_p\sin i\left(\frac{P_b}{2\pi}\right)^{-2/3}T_\odot^{-1/3}M^{2/3}M_c^{-1} = \sin i\,, \qquad\qquad (3.2)$$

might be precisely known from observations [18]. Here the total mass is M. The post-Keplerian parameters are functions of masses of the pulsar and its companion and Keplerian parameters. Any two post-Keplerian parameters determine the masses of both components very accurately. The best example of this is the double pulsar system PSR J0737-3039 [21]. Besides the five post-Keplerian parameters, the sixth parameter in the double pulsar system is extracted from the impact of the geodetic precession on the pulsar emission. Masses of pulsar A and B in PSR J0737-3039 are $1.3381 \pm 0.0007\,M_\odot$ and $1.2489 \pm 0.0007\,M_\odot$, respectively [22]. The determination of the post-Keplerian parameters in the Hulse–Taylor binary pulsar PSR B1913+16, PSR J1614-2230 and PSR J0740+6620 led to the accurate estimation of the mass of the pulsar in each case as $1.4398 \pm 0.0002\,M_\odot$, $1.97 \pm 0.04\,M_\odot$ and $2.14^{+0.10}_{-0.09}$ $M_\odot$, respectively [17, 23–25]. In the last two cases, this was estimated using the measurement of Shapiro delay, whereas $\dot{P}_b$, $\dot{\omega}$ and γ were measured in case of the Hulse–Taylor pulsar. The mass of PSR J0740+6620 was recently updated to a value of $2.08 \pm 0.07 M_\odot$ [26]. The other pulsars heavier than 2 $M_\odot$ are PSR J0348+0432 with $2.01 \pm 0.04\,M_\odot$ and PSR J1810+1744 with $2.13 \pm 0.04\,M_\odot$ [27, 28]. In both cases, the mass determinations were possible due to the pulsar timing and observations of companions. PSR J1810+1744 is a black widow pulsar which eats up its companion and the pulsar mass is estimated at 3σ confidence [28]. Figure 3.1 shows the compilation of pulsar masses in various binaries[1] [17].

3.2.2 Radius

The radius of a neutron star cannot be determined as precisely as the mass of a pulsar. Efforts were made to estimate the radius from thermal emissions from the surface of neutron stars. In this case, the radius measurement is based on the assumption of blackbody radiation. The measured flux with respect to an observer at an infinite distance is given by $F_\infty = \sigma T_{\text{inf}}^4 R_\infty^2/D^2$. Here T_∞ is the red-shifted surface temperature, D is the distance from the observer to the star and σ is the Stefan–Boltzmann constant. The radius that is determined in this fashion is not the actual radius but the radiation radius given by $R_\infty = R/\sqrt{1-2GM/R}$. The measurements of radii for several isolated neutron stars were carried out in this fashion. The isolated and nearby neutron star RX J1856-3784 was of the central

---

[1] https://stellarcollapse.org/nsmasses.

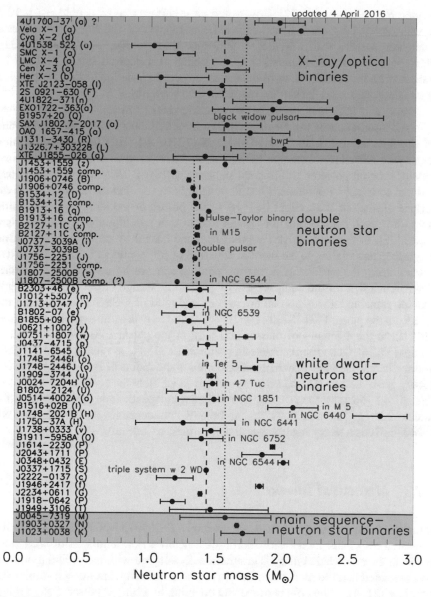

Fig. 3.1 Pulsar masses in binaries are shown here [17]

attraction in this context [29]. This neutron star was studied using the Hubble Space Telescope (HST) and ROSAT and Chandra x-ray telescopes. A distance of $D = 120 \pm 8\,\text{pc}$ was measured for this neutron star using the trigonometric parallax [30]. The thermal spectra of this source from x-ray to optical wavelengths were analyzed by different groups [17]. Nonmagnetic heavy element atmospheric models yielded $R_\infty \sim 16$ km [31, 32]. However, a large surface magnetic field $\sim 5 \times 10^{12}$ G was reported by other observations. Later models of the highly magnetized condensed surface with a hydrogen atmosphere led to $R_\infty > 13$ km [33, 34]. This shows that the determination of the radius of a neutron star is plagued by uncertainties in the estimation of the distance, poor knowledge of the composition of its atmosphere, and the surface magnetic field strengths.

A new and important campaign to measure radii of neutron stars has started with the installation of the Neutron Star Interior Composition Explorer (NICER) on the International Space Station since 2017 [35–37]. This telescope is geared toward the detection of thermal emissions in the soft x-rays band from the polar caps of rotation powered millisecond pulsars (MSPs). T. Strohmayer argued long ago that coherent x-ray intensity oscillations or burst oscillations originating on the surface of neutron stars might be used as the probes of the structures of neutron stars [38]. Such burst oscillations occur due to spin modulations of thermal burst fluxes [38]. In the case of MSPs, polar caps are heated by currents in the pulsar magnetosphere leading to the thermal emission of soft x-rays [35, 37]. The emitted photons from the surface of a rotating neutron star are influenced by the strong gravitational field of the compact star [38]. The knowledge of pulse profiles could provide valuable information on the masses and radii of MSPs. The NICER x-ray data from the pulsar PSR J0030+0451 were analyzed using pulse profile modeling. This led to the estimates of masses and radii of the pulsar independently by two groups [35, 36]. One group estimated a radius of $12.71^{+1.14}_{-1.19}$ km for the $1.34^{+0.15}_{-0.16}$ M$_\odot$ pulsar whereas the other group obtained the mass and radius of PSR J0030+0451 as $1.44^{+0.15}_{-0.14}$ M$_\odot$ and $13.02^{+1.24}_{-1.06}$ km, respectively [35, 36]. In the second observation of PSR J0740+6620 by the NICER, two different independent analyses reported the radius $12.39^{+1.30}_{-0.98}$ km and $13.71^{+2.61}_{-1.50}$ at the 68% confidence level [39, 40]. The PSR J0740+6620 is a heavy neutron star having a mass of 2.08 ± 0.07 M$_\odot$ [26].

3.2.3   Moment of Inertia

Relativistic binary pulsar systems, for example, PSR J0737-3039, hold key to the measurement of the moment of inertia of a neutron star. The spin of a compact object in the relativistic binary is gravitationally coupled with the orbital motion. It was predicted that the moment of inertia can be estimated from the spin-orbit (SO) coupling [21, 41]. The effects of the SO coupling in a binary pulsar system might be observed in the precession of the orbital plane and an extra contribution to the advance of the periastron. In the first case, the angular momentum (**L**) of the orbital plane precesses around the total angular momentum (**J**). This is also known as the geodetic precession. There would be compensating precessions of spins (**S$_A$** and **S$_B$**) of binary components as the total angular momentum (**J** = **L** + **S$_A$** + **S$_B$**) is conserved. The spin precession might impact the timing of binary pulsars [42]. As the orbital plane changes orientation, this would induce a change in the inclination angle which, in turn, might influence the time of arrival of pulses of a pulsar in the double pulsar system. The departure from the expected time of arrival of pulses depends sensitively how much the spin of the pulsar is misaligned with respect to the orbital angular momentum vector [41]. So far, this misalignment is found to be negligible from observations of pulsar A in PSR J0737-3039 [43]. Higher order post-Newtonian (PN) contributions could be determined in relativistic binary systems [19, 42]. The SO coupling may manifest in an extra advancement of the periastron above the PN effects. The total advance of the periastron is $\dot{\omega} = \dot{\omega}_{1PN} + \dot{\omega}_{2PN} + \dot{\omega}_{SO}$ [42]. According to Eq. (5.23) of Ref.[42], the expression of total advance of the periastron is,

$$\dot{\omega} = \frac{3\Gamma_0^2}{1 - e_T^2} \frac{2\pi}{P_b} \left[1 + d_0 \Gamma_0^2 - g_{S_A} \Gamma_0 \Gamma_{S_A} - g_{S_B} \Gamma_0 \Gamma_{S_B} \right], \qquad (3.3)$$

where P_b is the orbital period, $\Gamma_0 = \frac{(GM2\pi/P_b)^{1/3}}{c}$ denotes the ratio of the orbital velocity to the speed of light, $\Gamma_{S_i} = \frac{cI_i\Omega_i}{GM_i^2}$ represents the ratio of the spin velocity to the speed of light for the $i (= A, B)$-th binary component, I_i and Ω_i are the moment of inertia and angular spin frequency of i-th binary component, respectively, and e_T is the proper time eccentricity [44]. In Eq. (3.3), the first and second term represent the 1PN and 2PN contributions to $\dot{\omega}$, respectively, and the third and fourth terms imply the SO contributions due to spins $\mathbf{S_A}$ and $\mathbf{S_B}$, respectively. As the spin of pulsar B ($\mathbf{S_B}$) in PSR J0737-3039 is small, the fourth term can be neglected. The quantities d_0 and g_{S_A} are given by [19, 42, 43],

$$d_0 = \frac{1}{1 - e_T^2} \left(\frac{3}{2} x_A^2 + \frac{3}{2} x_A + \frac{27}{4} \right) - \left(\frac{5}{6} x_A^2 + \frac{23}{6} x_A - \frac{1}{4} \right),$$

$$g_{S_A}^{\|} = \frac{1}{(1 - e_T^2)^{1/2}} \left[\frac{1}{3} x_A^2 + x_A \right], \qquad (3.4)$$

with $x_A = M_A/M$, M_A and M are the mass of the pulsar A and the total mass of the binary, respectively. Here "$\|$" denotes that $\mathbf{L}$ and $\mathbf{S_A}$ are parallel [42]. It is noted that the SO coupling term in Eq. (3.3) depends on the moment of inertia (I).

The SO coupling is a tiny effect in $\dot{\omega}$ and could be significant when it is comparable to the 2PN contribution. The measurement of the SO effect would lead to the determination of the moment of inertia of the pulsar A in the double pulsar system. With the advent of the Square Kilometer Array (SKA) radio telescope, the number of detections of relativistic binary pulsar systems will increase significantly. Furthermore, dimensionally moment of inertia (I) $\propto MR^2$. In relativistic binaries, the mass of a neutron star is known precisely. Therefore, the determination of I leads to an estimate of the radius of a neutron star.

3.3 Compositions and Novel Phases of Neutron Stars—Crust to Core

The hot and neutrino-trapped protoneutron star born in core collapse supernova explosions cools down to temperatures of a few MeV in 20 s or so by emitting neutrinos. The matter in neutron stars is in the ground state and satisfies the conditions of the charge neutrality and chemical equilibrium under weak interactions. Compositions of neutron star matter differ from one layer to the other in its interior as density increases. It is inferred that the densest form of matter might be found in the central region of a neutron star. This matter is degenerate and the baryon chemical potential of a few hundred MeV is much higher than that of the temperature. It can be treated as a cold matter. This kind of cold and dense matter is beyond the reach of experiments in terrestrial laboratories. However, the Facility for Antiproton and Ion Research (FAIR) at GSI will investigate hot and dense matter formed in relativistic heavy ion collisions that might enrich our knowledge in understanding the gross properties of dense matter in a protoneutron star [45].

Broadly, the neutron star interior can be divided into two regions—the crust and core. The crust is again separated into the outer and inner crusts. Similarly this applies to the core with the outer and inner cores. The crust is made of matter at the sub-saturation density ($\leq 10^{14}$ g/cm^3). The matter in the crust is a nonhomogeneous mixture of nuclei, unbound neutrons and electrons. The full extent of the crust is $\sim$1–2 km. The crust contributes a few percent to the total mass of a neutron star. The knowledge of the crustal composition is important to understand the thermal and magnetic field evolution in neutron stars. The solid nature of the crust is essential in explaining glitches that are sudden spin-ups in pulsars.

The outer crust begins at a density $\sim 10^4$ g/cm^3. Here ionized nuclei are immersed in a degenerate electron gas. Nuclei are arranged in a body-centered cubic (bcc) lattice. The equilibrium nucleus at this density is ^{56}Fe. As density increases, the equilibrium nucleus becomes more and more neutron-rich. On the other hand, electrons become relativistic above densities $> 10^7$ g/cm^3. Some neutrons are no longer bound in a nucleus when the density reaches the value $\sim$4–7 $\times$ 10^{11} g/cm^3 and those drip out of the nucleus. In this situation, the neutron chemical potential exceeds the rest mass of neutrons. The last equilibrium nucleus in this region is ^{118}Kr. The neutron drip point marks the end of the outer crust and the beginning of the inner crust.

The matter in the inner crust is comprised of neutron-rich nuclei and free neutrons in a background of relativistic electrons. Free neutrons form Cooper pairs due to the attractive nuclear interaction and result in superfluidity when the neutron star temperature is below the critical temperature. In the layer of matter in close proximity to the core, nuclei undergo shape transitions from spherical to cylindrical, slab, cylindrical bubble, and spherical bubble with increasing density. This is known as the pasta phase. Nuclei dissolve into their fundamental constituents of neutrons and protons and a uniform nuclear matter is formed at the saturation density $\sim$2.7 $\times$ 10^{14} g/cm^3. It is noted that the electron chemical potential increases with the baryon density.

The outer core extends up to twice the saturation density. Mostly neutrons along with equal fractions of protons and electrons are the main ingredients of the uniform matter here. Muons might appear when the electron chemical potential is equal to or greater than the bare mass of muons. Electrons and muons are treated as noninteracting particles. The depth of the outer core could be a few kilometers.

The inner core is the central region of a neutron star. The density in this regime might exceed by several times the saturation density. As the density of matter is very high, nucleons at the Fermi energy levels decay into heavier baryons subject to charge neutrality, baryon number conservation, and weak chemical equilibrium conditions. This process would also lead to the population of strange baryons [46–48]. The strangeness conservation is violated in weak interaction processes. Consequently, a net strangeness will develop in neutron star matter. Other novel forms of (non)strange matter such as Δ baryons might appear in neutron star interior [49]. It was predicted that neutrons could decay into mesons, for example, pions and kaons in dense hadronic matter produced in heavy ion collisions and form Bose–Einstein condensates [50–52]. The kaon condensation was later explored in neutron star matter [53–64]. Hadrons dissolve into their fundamental constituents at very high densities and lead to the creation of the (un)paired strange quark matter [65–69]. The dense matter in the core is mainly responsible in determining the masses of neutron stars.

3.4 Equation of State Models of Neutron Star Matter

A neutron star can be thought of as a "giant nucleus." The physics of dense star matter in it is a many-body problem. There are different methods to deal with the many-body problem of homogeneous and inhomogeneous matter in neutron stars. Any relation between thermodynamic state variables is defined as the equation of state (EoS). The EoS models of neutron star matter suffer from uncertainties such as the inadequate knowledge of interactions and the treatment of the many-body problem. These uncertainties of an EoS model might be constrained using terrestrial experimental and observational data [70]. It is a challenge for us to formulate techniques that describe new physics associated with the dense matter far away from the normal nuclear matter. In the following paragraphs, we discuss different models of strongly interacting matter that capture the essence of many-body systems.

The EoS models of beta-equilibrated and charge neutral homogeneous matter in neutron stars can be broadly divided into non-relativistic and relativistic models. Furthermore, these models can be classified as (i) microscopic models, (ii) chiral effective field theory models, and (iii) phenomenological models.

3.4.1 Microscopic Models

The widely used many-body methods under this category are non-relativistic Brueckner–Hartree–Fock (BHF), variational calculations, and the relativistic Dirac-Brueckner–Hartree-Fock (DBHF) calculation [71–75]. The starting points of these many-body techniques are two and three-body interactions whose parameters are fitted to nucleon-nucleon scattering data in vacuum and the properties of the deuteron.

The main difficulty in formulating the nuclear many-body theory based on a realistic nucleon-nucleon interaction is the strong repulsive core at the short-range. The many-body approach developed by Brueckner, Bethe, Goldstone and others took care of the strong repulsive core in the Brueckner theory of nuclear matter. The short-range correlations among pairs of nucleons in the Brueckner-Bethe-Goldstone (BBG) formalism are included in the G matrix [71, 72, 76],

$$G = \mathcal{V} - \mathcal{V}\frac{Q}{e}G \,, \tag{3.5}$$

where $\mathcal{V}$, Q, and e are the bare nucleon-nucleon potential, the Pauli exclusion operator and the energy denominator in the Goldstone diagram, respectively. The G matrix represents the effective two-body interaction and is derived perturbatively by summing all ladder diagrams. The BHF formalism could not reproduce the empirical values of the saturation density and binding energy of the nuclear matter. However, the BHF calculations with the three-body interaction are found to improve the situation to some extent [76]. However, this is not enough to be in good agreement with the saturation properties.

The Brueckner theory of nuclear matter was extended to the relativistic case known as the Dirac-Brueckner (DB) formalism [73, 77–80]. Like the non-relativistic Brueckner theory, a G matrix is the main ingredient in the relativistic case. The salient features of the Dirac-Brueckner approach is the use of the Dirac equation to describe the single particle motion in nuclear matter. The free nucleon-nucleon interaction adopted in the Dirac-Brueckner calculations is a one boson exchange nucleon-nucleon potential. The nucleon self-energy in this relativistic approach is made of attractive scalar and repulsive vector fields. The saturation properties of nuclear matter were calculated in the Dirac-Brueckner–Hartree-Fock approximation and agreed reasonably well with the empirical values. The success of the DBHF approximation may be attributed to the dressing of the in-medium spinors which introduces a density dependent interaction.

The many-body problem of nuclear matter was also investigated in the non-relativistic variational approach. The starting point in this approach is a Hamiltonian involving nuclear potentials constructed particularly for variational calculations. A specific example of it is the Argonne nucleon-nucleon interactions. Furthermore, a three-body interaction was added to the Hamiltonian such as the Urbana three-body interactions. A variational trial wave function is constructed with correlations included in it as,

$$\Psi_{trial}(1, \dots, N) = F(1, \dots, N)\Phi(1, \dots, N) , \qquad (3.6)$$

where the operator F transforms the uncorrelated and antisymmetrized wave function of N single particle wave functions. The ground state energy of the many-body system is obtained from the expectation value of the nuclear Hamiltonian using the trial wave function. This gives an upper bound on the ground state energy.

The main focus of the variational approach is to find out an ansatz for the correlation operator F which might have the same form as the two-body interaction [70, 81–83]. The correlation operator is constructed as a sum of the product of radial correlation functions and spin-isospin dependent two-body operators [70, 75]. This might be realized using different techniques such as variational chain summation, Fermi hypernetted chain, coupled cluster techniques [75, 84–88]. The calculation of Akmal, Pandharipande, and Ravenhall (APR) is the most celebrated work on the dense matter in neutron stars in this category [75].

3.4.2 Chiral Effective Field Theory Models

The quantum chromodynamics (QCD) is the fundamental theory to describe the ground state matter in neutron star cores [89–91]. A perturbative approach within the QCD framework is best suited for the highly dense matter due to the weak interactions among quarks. However, the ground state of nuclear matter made of nucleons and mesons is a highly non-perturbative system. This situation could be treated within the effective field theory (EFT).

An approximate symmetry of the low-energy QCD is the chiral symmetry which is spontaneously broken. The chiral effective field theory (ChEFT) appears as an elegant tool to describe the nuclear matter at the saturation density and up to densities not far off from the saturation point [92–94]. The relevant degrees of freedom in this theory are nucleons and pions that are Goldstone bosons of the spontaneously broken symmetry. The ChEFT is based on the systematic expansion of two-body, three-body, and many-body nuclear interactions within the theory's range of validity. The Lagrangian of the ChEFT is the most general one which is consistent with the symmetries of low-energy including the spontaneously broken chiral symmetry. The power counting scheme which is given by the powers of a typical momentum or the pion mass over the chiral breakdown scale is exploited to organize the most important contributions among a very large number of Feynman diagrams. The theoretical uncertainties within this approach are computed by the inspection of the order by order convergence of the ChEFT expansion. This model is being used widely for the study of dense matter in neutron stars [94, 95].

3.4.3   Phenomenological Models

These approaches include both non-relativistic and relativistic models. We discuss several such models being used for the description of neutron star matter.

In the non-relativistic models, the effective nucleon-nucleon interactions are used instead of realistic nuclear forces. The effective interaction is exploited to calculate the properties of finite nuclei as well as those of nuclear matter. The most prominent among the effective interactions is the Skyrme interaction [96–100]. The Skyrme interaction is a zero range interaction. It also includes an explicit three-body interaction. The other widely used effective interactions are the Gogny interaction and modified Seyler-Blanchard interaction [101–103]. Both are finite-range effective interactions and involve density dependent effective two-body interactions. However, the non-relativistic phenomenological approaches suffer a set back at high densities because the EoS becomes superluminal [102].

The answer to the superluminal behavior of the EoS at high densities is the relativistic field theoretical model. Such a model based on the baryon and meson degrees of freedom is attractive on several accounts. A microscopic treatment of the many-body nuclear matter problem in this model respects the special theory relativity, causality, and the symmetries such as Lorentz covariance, parity conservation, isospin symmetry. The relativistic field theory of hadrons is known as the quantum hadrodynamics (QHD) [104–107]. The QHD is characterized by coupling constants that are calibrated to observed nuclear properties and could be extrapolated into extreme astrophysical conditions of high density and temperature. The relativistic field theory of hadrons was successful in describing gross properties of nuclei and nuclear matter.

There were important efforts to construct a quantum hadron field theory of nuclear interactions from a microscopic approach. The motivation behind such an approach is to retain the essential features of the QHD [106] as well as to deal with the complicated many-body dynamics of strong interactions [108–110]. An appropriate and successful microscopic approach to the in-medium nuclear interactions follows from the Dirac-Brueckner calculations. The DB calculations with realistic nucleon-nucleon interactions reproduced empirical saturation properties of symmetric nuclear matter reasonably well and calculations for the nucleons-only neutron star matter were also done within this formalism [79, 111–119]. The bulk of the screening of nucleon-nucleon interaction in the medium was taken into account by the local baryon density dependent DB self-energies making the relativistic many-body dynamics to be approximated by a density dependent relativistic hadron (DDRH) field theory [109, 110].

A covariant and thermodynamically consistent DDRH field theory is derived by making interaction vertices as Lorentz scalar functionals of baryon field operators. In the mean field approximation, this model becomes the relativistic Hartree description with density dependent meson-nucleon couplings. The density dependent meson-nucleon couplings were obtained from the Dirac-Brueckner–Hartree-Fock nucleon self-energies calculated with Bonn, Groningen and phenomenological density dependent potentials [120–125]. The functional forms of the density dependent couplings were also introduced in the parametrization of the density dependent relativistic mean field (RMF) model that fitted to the nuclear binding energies [122]. The derivatives of vertices with respect to baryon fields lead to rearrangement terms in baryon field equations [110]. The DDRH model without rearrangement terms was first used to study finite nuclei [108]. Later the DDRH model with the rearrangement terms was adopted to investigate deformed nuclei [121], hypernuclei [123], asymmetric nuclear matter, exotic nuclei [124, 125] as well as neutron star properties [126].

3.4.4 EoS Models of Matter at Sub-saturation Density

The matter at sub-saturation densities is inhomogeneous. This matter is made of light and heavy nuclei along with unbound neutrons. In the mass density regime 10^4–$\sim 10^{14}$ g/cm^3, the spatial region comprised of the inhomogeneous nuclear matter in a uniform background of electrons as well as in β equilibrium makes the crust of neutron stars. The EoS of the crust is matched with that of the uniform matter in the core to construct an overall EoS model for the calculation of gross properties of neutron stars. The crust EoS plays an important role in the determination of the radii of neutron stars. Many studies were carried out to describe the inhomogeneous matter in the crust at zero and finite temperatures [70]. Here we discuss some of those widely used crust models.

3.4.4.1 Baym–Pethick–Sutherland Model of Outer Crust

The pioneering work of Baym, Pethick, and Sutherland (BPS) focused on the equilibrium composition and EoS of the ground state matter of the outer crust [128]. The Gibbs free energy per baryon (g) was minimized with respect to mass (A) and atomic (Z) numbers at a constant pressure (P) to get the equilibrium sequence of nuclei [128]. The Gibbs energy per nucleon is given by,

$$g = \frac{E_{tot} + P}{n_B} = \frac{W_N + 4/3 W_L + Z\mu_e}{A},$$
$$E_{tot} = n_N(W_N + W_L) + \varepsilon_e(n_e),$$
$$P = P_e + \frac{1}{3} W_L n_N, \tag{3.7}$$

where the baryon number density, n_B, and the number density of nuclei, n_N, are connected by $n_B = A n_N$. The energy of the nucleus including the rest mass energy of nucleons is given by

$$W_N = m_n(A - Z) + m_p Z - bA, \tag{3.8}$$

where b is the binding energy per nucleon. Experimental nuclear masses are taken from the atomic mass table of Audi et al. [129] and we adopt the theoretical extrapolation for the rest of nuclei [130]. The lattice energy of the cell (W_L) is,

$$W_L = -\frac{9}{10}\frac{Z^2 e^2}{r_C}\left(1 - \frac{5}{9}\left(\frac{r_N}{r_C}\right)^2\right). \qquad (3.9)$$

The cell and nucleus radii are r_C and r_N, respectively. The first and second terms in Eq. (3.9) represent the lattice energy for the point nucleus and the finite size effect, respectively.

In Fig. 3.2, the Gibbs energy per particle for the sequence of equilibrium nuclei is plotted as a function of mass density. The equilibrium nucleus at the lower density in the outer crust is populated by ^{56}Fe. As density increases, nuclei become more and more neutron-rich. The outer crust ends with ^{118}Kr after which neutrons drip out of the nucleus and the inner crust begins.

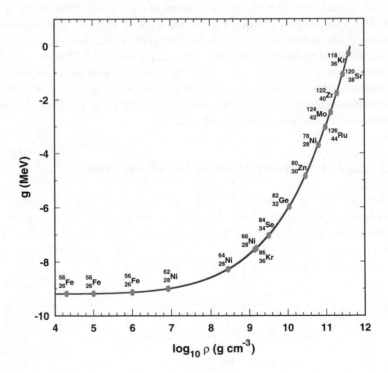

Fig. 3.2 Gibbs energy per nucleon is shown as a function of mass density for the sequence of equilibrium nuclei in the outer crust. This is taken from Ref. [127] and reprinted by permission from Springer Nature

3.4.4.2 EoS Models of Inner Crust

The β-equilibrated and charge neutral matter of the inner crust is composed of nuclei embedded in a neutron gas and the background of an uniform electron gas. Nuclei are considered to be in mechanical equilibrium with the neutron gas. The semiempirical mass formula was extrapolated to the free neutron gas regime in the early studies of the inner crust [131, 132]. The reduction of the nuclear surface energy due to the free neutron gas was considered in the work of Baym, Bethe, and Pethick (BBP) [133]. The appearance of unusual nuclear shapes with increasing density in the inner crust might be possible [134, 135]. The investigations of nuclei in the inner crust were also carried out using the relativistic field theoretical model [136].

J.W. Negele and D. Vautherin investigated nuclear clusters in the neutron star crust using the energy density of a many-body system [137]. Their many-body nuclear theory was derived from the two-body effective nuclear interaction taking into account two-body correlations in it. The energy density of the many-body nuclear matter was constructed as simple functionals of number density and kinetic energy density exploiting the density matrix expansion. Eleven ground state nuclear configurations in the free neutron regime relevant to the inner crust were investigated and the EoS was obtained fitting those configurations at eleven discrete densities by Negele and Vautherin [137]. Nuclear clusters were labelled by the total number of nucleons in the Wigner-Seitz unit cell as neutrons in nuclei could not be separated from those in the gas. It was noted that the EoS of Negele and Vautherin had a good agreement with that of the BBP EoS.

There are important aspects in the problem of nuclei in a neutron gas. Firstly the coexistence of two phases of nuclear matter—denser phase inside a nucleus and low density phase outside it, is to be dealt with in a thermodynamically consistent manner. Secondly, the surface energy of the interface between two phases should be determined with good accuracy. This problem could be solved within the framework of the temperature dependent Hartree-Fock theory adopting the subtraction procedure of Bonche, Levit, and Vautherin (BLV) [138–140]. This same method was extended to the study of isolated nuclei embedded in a neutron gas [141] as well as nuclei in the inner crust at zero temperature in the Thomas-Fermi (TF) approximation [142]. In this case, nuclei arranged in a lattice, reside in a nucleonic gas as well as a uniform background of electrons. In the Wigner-Seitz (WS) approximation, each lattice volume is replaced by a spherical cell. Each cell is charge neutral with electrons uniformly distributed within it and there is no Coulomb interaction between cells. The system is in the β-equilibrium.

In the WS cell, a nucleus is located at the center surrounded by a low density neutron gas whereas protons are part of the nucleus. As the spherical cell does not represent a nucleus, the nucleus is extracted after the subtraction of the gas part from the cell following the prescription of Bonche, Levit, and Vautherin [138, 139]. It was also demonstrated that the TF formalism at finite temperatures gave two solutions—one of those corresponds to the nucleus plus neutron gas and the other solution represents only the neutron gas. The density profiles of the nucleus plus neutron gas and that of the neutron gas can be determined self-consistently in the TF formalism. Consequently, the nucleus is obtained as the difference of two solutions. This formalism applied to the calculation at zero temperature is described below.

The definition of the thermodynamic potential (Ω_N) of the nucleus is [138, 139]

$$\Omega_N = \Omega_{NG} - \Omega_G , \tag{3.10}$$

and

$$\Omega = \mathcal{F} - \sum_{q=n,p} \mu_q A_q , \tag{3.11}$$

where Ω_{NG} is the thermodynamic potential of the nucleus plus gas phase (NG) and Ω_G is that of the gas (G) only, μ_q and A_q are the chemical potential and number of q-th nucleon, respectively. The nucleus plus gas solution coincides with the gas solution at large distance. The free energy

$$\mathcal{F}(n_B, Y_p, T) = \int [\mathcal{H} - Ts + \varepsilon_c + f_e] d\mathbf{r} \tag{3.12}$$

is expressed in terms of the nuclear energy density functional $\mathcal{H}$, the entropy density of nucleons s, the Coulomb energy density ε_c, and the free energy density of electrons $f_e = \varepsilon_e - Ts_e$. The free energy depends on the average baryon density (n_B), proton fraction (Y_p) and temperature (T). We study the properties of nuclei of the inner crust at $T = 0$.

The nuclear energy density based on the SkM nucleon-nucleon interaction is given by [142–144]

$$
\begin{aligned}
\mathcal{H}(r) = {} & \frac{\hbar^2}{2m_n^*} \tau_n + \frac{\hbar^2}{2m_p^*} \tau_p + \frac{1}{2} t_0 \left[\left(1 + \frac{x_0}{2}\right) n_B^2 - \left(x_0 + \frac{1}{2}\right) \left(n_n^2 + n_p^2\right) \right] \\
& - \frac{1}{16} \left[t_2 \left(1 + \frac{x_2}{2}\right) - 3t_1 \left(1 + \frac{x_1}{2}\right) \right] (\nabla n_B)^2 \\
& - \frac{1}{16} \left[3t_1 \left(x_1 + \frac{1}{2}\right) + t_2 \left(x_2 + \frac{1}{2}\right) \right] \left[(\nabla n_n)^2 + (\nabla n_p)^2\right] \\
& + \frac{1}{12} t_3 n_B^\alpha \left[\left(1 + \frac{x_3}{2}\right) n_B^2 - \left(x_3 + \frac{1}{2}\right) \left(n_n^2 + n_p^2\right) \right] ,
\end{aligned}
\tag{3.13}
$$

where $x_0, x_1, x_2, x_3, t_1, t_2, t_3$ are parameters of the interactions. The first and second terms are kinetic energy densities of neutrons and protons, respectively. This is followed by the contributions from the zero range part of the Skyrme interaction, surface effects and the density dependent part of the nucleon-nucleon interaction.

The effective mass of nucleons is given by

$$\frac{m}{m_q^*(r)} = 1 + \frac{m}{2\hbar^2} \left\{ \left[t_1 \left(1 + \frac{x_1}{2} \right) + t_2 \left(1 + \frac{x_2}{2} \right) \right] n_B \right.$$

$$\left. + \left[t_2 \left(x_2 + \frac{1}{2} \right) - t_1 \left(x_1 + \frac{1}{2} \right) \right] n_q \right\}, \tag{3.14}$$

where total baryon density is $n_B = n_n + n_p$.

The Coulomb energy densities for the NG and G phases are:

$$\varepsilon_c^{NG}(r) = \frac{1}{2}(n_p^{NG}(r) - n_e) \int \frac{e^2}{|\mathbf{r} - \mathbf{r}'|}(n_p^{NG}(r') - n_e)d\mathbf{r}'$$

$$\varepsilon_c^{G}(r) = \frac{1}{2}(n_p^{G}(r) - n_e) \int \frac{e^2}{|\mathbf{r} - \mathbf{r}'|}(n_p^{G}(r') - n_e)d\mathbf{r}'$$

$$+ n_p^{NG}(r) \int \frac{e^2}{|\mathbf{r} - \mathbf{r}'|}(n_p^{G}(r') - n_e)d\mathbf{r}', \tag{3.15}$$

where n_p^{NG} and n_p^{G} are proton densities in the nucleus plus gas and only gas phase. Here the direct part of coulomb energy densities (ε_c) is shown as the contribution of the exchange part is small.

The thermodynamic potential is minimized subject to the condition of number conservation of each species (neutron or proton) and the density profiles of neutrons and protons are computed from the following relations

$$\frac{\delta \Omega_{NG}}{\delta n_q^{NG}} = 0 \,,$$

$$\frac{\delta \Omega_{G}}{\delta n_q^{G}} = 0 \,. \tag{3.16}$$

This leads to the following coupled equations [141, 142]

$$(3\pi^2)^{\frac{2}{3}} \frac{\hbar^2}{2m_q^*}(n_q^{NG})^{\frac{2}{3}} + V_q^{NG} + V_c^{NG}(n_p^{NG}, n_e) = \mu_q \,,$$

$$(3\pi^2)^{\frac{2}{3}} \frac{\hbar^2}{2m_q^*}(n_q^{G})^{\frac{2}{3}} + V_q^{G} + V_c^{G}(n_e) = \mu_q \,, \tag{3.17}$$

with the effective mass of the q-th species m_q^*, the single particle potentials of nucleons in the nucleus plus gas and gas phases V_{NG}^q and V_G^q [143]. On the other hand, direct parts of the single particle Coulomb potential corresponding to the nucleus plus gas (V_c^{NG}) and only gas (V_c^{G}) solutions are given by

$$V_c(r) = \int \left[n_p^{NG}(r') - n_e \right] \frac{e^2}{|\mathbf{r} - \mathbf{r}'|} d\mathbf{r}' . \tag{3.18}$$

The average chemical potential for q-th nucleon is

$$\mu_q = \frac{1}{A_q} \int \left[(3\pi^2)^{\frac{2}{3}} \frac{\hbar^2}{2m_q^*} (n_q^{NG})^{\frac{2}{3}} + V_q^{NG}(r) + V_c^{NG}(r) \right] n_q^{NG}(r) d\mathbf{r} . \tag{3.19}$$

Here A_q denotes either N_{cell} or Z_{cell} of the cell defined by the average baryon density n_B and proton fraction Y_p. Furthermore, the β-equilibrium condition is

$$\mu_n = \mu_p + \mu_e . \tag{3.20}$$

Density profiles of neutrons and protons in the cell are constrained with the neutron number (N_{cell}) and proton number (Z_{cell}) in the cell as

$$Z_{cell} = \int n_p^{NG}(r) d\mathbf{r} ,$$

$$N_{cell} = \int n_n^{NG}(r) d\mathbf{r} . \tag{3.21}$$

Finally, the subtraction procedure is implemented to get the number of neutrons (N) and protons (Z) in a nucleus with the mass number $A = N + Z$ using

$$Z = \int \left[n_p^{NG}(r) - n_p^{G}(r) \right] d\mathbf{r} ,$$

$$N = \int \left[n_n^{NG}(r) - n_n^{G}(r) \right] d\mathbf{r} . \tag{3.22}$$

The free energy per nucleon is the sum of the nuclear energy (e_N) including the Coulomb interaction among protons, the lattice energy (e_{lat}) that involves the interaction between electrons and protons and the electron kinetic energy (e_{ele}) and given by

$$F/A = e_N + e_{lat} + e_{ele}. \tag{3.23}$$

The different contributions in the free energy compete with each other to make it minimum.

Nuclear clusters and the EoS of the inner crust are calculated at zero temperature. The energy per particle (E/A) for the sequence of nuclear clusters which include nuclei along with neutrons in the gas phase is exhibited as a function of mass density in Fig. 3.3. The starting nucleus of the inner crust is ^{125}Kr in which neutrons

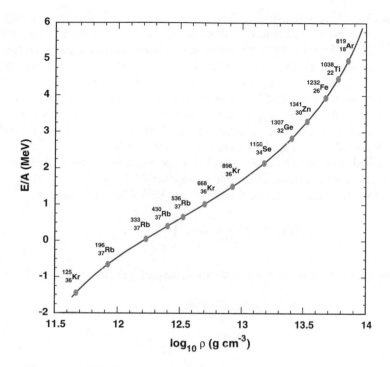

Fig. 3.3 Energy per nucleon is shown as a function of mass density for nuclear clusters of the inner crust

just begin to drip out. As the mass density increases, the number of neutrons in nuclear clusters increases reaching a maximum followed by the decrease in neutron numbers. A similar trend was noted in the calculation of nuclear clusters by Negele and Vautherin [137].

As we compute the density profiles of neutrons and protons in the nucleus plus gas as well as gas phases at each baryon density, we obtain the neutron number, the proton number, and the mass number of the nucleus using the subtraction procedure as in Eq. (3.22). Calculations were performed with the Skyrme interactions SkM and SLy4 [145]. It was noted that the mass number increased with baryon density, reached a maximum, and fell afterwards. On the other hand, the proton number decreased from a maximum value with the increasing baryon density.

3.4.4.3 Extended Nuclear Statistical Equilibrium Model

We discuss the extended Nuclear Statistical Equilibrium (NSE) model to describe the inhomogeneous matter including light and heavy nuclei along with unbound nucleons at low densities below the saturation density [146]. In this model, nuclei are described as non-relativistic particles obeying the Maxwell–Boltzmann statistics and medium corrections such as internal excitations or Coulomb screening. The dissolution of heavy nuclei at high densities is implemented through the excluded volume effects. Interactions among unbound nucleons are described by the relativistic field theoretical models that we discuss in the next section. Experimental masses of nuclei ($A \geq 2$) in the Hempel and Schaffner-Bielich (HS) model are sourced from the atomic mass table of [129]. Theoretical nuclear structure calculations within the framework of the finite-range droplet model (FRDM) are taken into account for exotic nuclei without measured masses [130]. The extended NSE model includes more than 8000 nuclear species down to the neutron drip line. Nuclear shell effects are automatically included in the calculation because nuclear mass tables are being used. Nuclei are only considered up to a temperature $\sim$50 MeV.

The total canonical partition function in the NSE model is written as

$$Z(T, V, \{N_i\}) = Z_{\text{nuc}} \prod_{A,Z} Z_{A,Z} \, Z_{\text{Coul}} , \tag{3.24}$$

and the Helmholtz free energy follows from the partition function as,

$$F(T, V, \{N_i\}) = -T \ln Z \tag{3.25}$$

$$= F_{\text{nuc}} + \sum_{A,Z} F_{A,Z} + F_{\text{Coul}} , \tag{3.26}$$

V denoting the volume of the system; F_{nuc}, F_{Coul}, $F_{A,Z}$ are the free energies of nucleons, the Coulomb free energy, and the free energy of the nucleus given by the Maxwell–Boltzmann distribution [146].

The excluded volume effects is introduced in a thermodynamically consistent manner in the model. The number densities of nuclei are calculated as [146],

$$n_{A,Z} = \kappa \, g_{A,Z}(T) \left(\frac{M_{A,Z} T}{2\pi} \right)^{3/2}$$

$$\times \exp \left(\frac{(A - Z)\mu_n^0 + Z\mu_p^0 - M_{A,Z} - E_{A,Z}^{\text{Coul}} - P_{\text{nuc}}^0 V_{A,Z}}{T} \right) , \tag{3.27}$$

where κ, the volume fraction available for nuclei defined in terms of local number densities takes values between 0 and 1. The internal partition function of nuclei, $g_{A,Z}(T)$ in Eq. (3.27), which is taken from [147], involves only excited states up to the binding energy of the corresponding nucleus to keep the nucleus bound. Arbitrarily large excitation energies would contribute to the energy density in the absence of a cutoff in the integral for the excited states. Consequently, the energy and entropy stored in nuclei would increase to unphysically large values with increasing temperature. It was noted in different applications of the EoS that the usage of the cutoff led to a more well-balanced behavior.

Next the free energy density is defined as [146]

$$f = \sum_{A,Z} f^0_{A,Z}(T, n_{A,Z}) + f_{\text{Coul}}(n_e, n_{A,Z}) + \xi f^0_{\text{nuc}}(T, n'_n, n'_p) - T \sum_{A,Z} n_{A,Z} \ln(\kappa) \,,$$

(3.28)

where the first term is the contribution of the noninteracting gas of nuclei and f_{Coul} is the Coulomb free energy density. The free energy density of the interacting nucleons is f^0_{nuc} and the available volume fraction of nucleons is ξ. The local number densities of neutrons and protons are labeled by n'_n and n'_p, respectively. The last term that takes care of the hard-core repulsion of nuclei, goes to infinity as κ approaches zero near the saturation density when the uniform matter is formed.

The energy density has the following expression in the extended NSE model [146],

$$\epsilon = \xi \epsilon^0_{\text{nuc}}(T, n'_n, n'_p) + \sum_{A,Z} \epsilon^0_{A,Z}(T, n_{A,Z}) + f_{\text{Coul}}(n_e, n_{A,Z}) \,,$$
(3.29)

where

$$\epsilon^0_{A,Z}(T, n_{A,Z}) = n_{A,Z} \left(M_{A,Z} + \frac{3}{2}T + \frac{\partial g}{\partial T} \frac{T^2}{g} \right) \,.$$
(3.30)

The total pressure is

$$P = P^0_{\text{nuc}}(T, n'_n, n'_p) + \frac{1}{\kappa} \sum_{A,Z} P^0_{A,Z}(T, n_{A,Z}) + P_{\text{Coul}}(n_e, n_{A,Z}) \,,$$
(3.31)

where

$$P^0_{A,Z}(T, n_{A,Z}) = T n_{A,Z} \,.$$
(3.32)

3.5   Relativistic Field Theoretical Models for Dense Matter at Zero and Finite Temperatures

Here we discuss the relativistic field theory models based on the Walecka model and its extension to describe the many-body hadronic matter in neutron stars at zero and finite temperatures. These calculations are performed in the flat spacetime because one makes a negligible error for not including the general theory of relativity [148].

3.5.1   Relativistic Mean Field Models

The Lagrangian density for baryon-baryon interaction mediated by the exchange of scalar σ, vector ω and isovector ρ mesons is given by [148],

$$
\mathcal{L}_B = \sum_{B=N,\Lambda,\Sigma,\Xi} \bar{\Psi}_B \left(i\gamma_\mu \partial^\mu - m_B + g_{\sigma B}\sigma - g_{\omega B}\gamma_\mu \omega^\mu - g_{\rho B}\gamma_\mu \boldsymbol{\tau}_B \cdot \boldsymbol{\rho}^\mu \right) \Psi_B
$$

$$
+ \frac{1}{2}\left(\partial_\mu \sigma \partial^\mu \sigma - m_\sigma^2 \sigma^2 \right) - U(\sigma)
$$

$$
- \frac{1}{4}\omega_{\mu\nu}\omega^{\mu\nu} + \frac{1}{2}m_\omega^2 \omega_\mu \omega^\mu - \frac{1}{4}\boldsymbol{\rho}_{\mu\nu} \cdot \boldsymbol{\rho}^{\mu\nu} + \frac{1}{2}m_\rho^2 \boldsymbol{\rho}_\mu \cdot \boldsymbol{\rho}^\mu . \tag{3.33}
$$

Dirac spinors, Ψ_B, denote the isospin multiplets for baryons $B = N, \Lambda, \Sigma, \Xi$, the bare baryon mass is m_B, and τ_B is the isospin operator. Meson-baryon coupling constants in Eq. (3.33) are $g_{\sigma B}$, $g_{\omega B}$, and $g_{\rho B}$. The scalar σ and vector ω meson terms generate the long-range attraction and short-range repulsion of the nuclear interaction, respectively. The field strength tensors are given by,

$$
\omega_{\mu\nu} = \partial_\mu \omega_\nu - \partial_\nu \omega_\mu ,
$$

$$
\rho_{\mu\nu} = \partial_\mu \rho_\nu - \partial_\nu \rho_\mu . \tag{3.34}
$$

The non-linear self-interactions of the scalar field added to get a realistic compression modulus are written as [149]

$$
U(\sigma) = \frac{1}{3}g_2 \sigma^3 + \frac{1}{4}g_3 \sigma^4 , \tag{3.35}
$$

where g_2 and g_3 are constants.

The Euler–Lagrange equation,

$$
\partial^\mu \left(\frac{\partial \mathcal{L}}{\partial (\partial^\mu q)} \right) - \frac{\partial \mathcal{L}}{\partial q} = 0 , \tag{3.36}
$$

when solved for each field (q), gives the equation of motion for the respective field as,

$$
\partial^\mu \partial_\mu \sigma + m_\sigma^2 \sigma = \sum_B g_{\sigma B} \bar{\psi}_B \psi_B - \frac{\partial U}{\partial \sigma} , \tag{3.37}
$$

$$
\partial^\nu \omega_{\mu\nu} + m_\omega^2 \omega_\mu = \sum_B g_{\omega B} \bar{\psi}_B \gamma_\mu \psi_B , \tag{3.38}
$$

$$
\partial^\nu \boldsymbol{\rho}_{\mu\nu} + m_\rho^2 \boldsymbol{\rho}_\mu = \sum_B g_{\rho B} \bar{\psi}_B \gamma_\mu \boldsymbol{\tau}_B \psi_B , \tag{3.39}
$$

and for baryon B

$$\left[i\gamma^\mu\partial_\mu - (m_B - g_{\sigma B}\sigma) - g_{\omega B}\gamma^\mu\omega_\mu - g_{\rho B}\gamma^\mu\boldsymbol{\tau}_B\cdot\boldsymbol{\rho}_\mu\right]\psi_B = 0 . \quad (3.40)$$

The exact solutions of the meson field equations are complicated because of the presence of source terms in those equations. Furthermore, the meson-baryon couplings are too large to implement a perturbative approach. It is to be noted that the source terms in the meson field equations become large as the baryon density increases. In this situation, the meson fields are replaced by their expectation values with respect to the ground state of the nuclear matter. This is known as the relativistic mean field (RMF) approximation. Here we are dealing with the many-body system of the nuclear matter in the ground state. The system is also uniform and stationary. This implies that the meson fields do not depend on the space and time. As the nuclear matter is at rest, the space parts of ω_μ and ρ_μ fields vanish. Moreover, the third component of the isovector ρ meson couples to baryon because the expectation values of the sources of the charged $\rho^\pm$ mesons vanish in the ground state of the system. Replacing the meson fields by their expectation values in the ground state as denoted by $\sigma \to \langle\sigma\rangle$, $\omega_0 \to \langle\omega_\mu\rangle$, and $\rho_{03} \to \langle\rho_{03}\rangle$, the equations of motion are rewritten as,

$$m_\sigma^2\sigma = \sum_B g_{\sigma B}\langle\bar\psi_B\psi_B\rangle ,$$

$$m_\omega^2\omega_0 = \sum_B g_{\omega B}\langle\bar\psi_B\gamma_0\psi_B\rangle ,$$

$$m_\rho^2\rho_{03} = \sum_B g_{\rho B}\langle\bar\psi_B\gamma_0\tau_{3B}\psi_B\rangle . \quad (3.41)$$

Further, the baryon field equation reduces to,

$$\left[i\gamma^\mu\partial_\mu - m_B^* - g_{\omega B}\gamma^0\omega_0 - g_{\rho B}\tau_{3B}\gamma^0\rho_{03}\right]\psi_B = 0 . \quad (3.42)$$

The effective mass of baryon B is defined as $m_B^* = m_B - g_{\sigma B}$.

The EoS of baryonic matter is a relation between the energy density and pressure, and derived from the energy-momentum tensor,

$$\Gamma_{\mu\nu} = -\eta_{\mu\nu}\mathcal{L} + \frac{\partial\mathcal{L}}{\partial(\partial^\mu\varphi)}(\partial_\nu\varphi) , \quad (3.43)$$

where Minkowski metric $\eta_{\mu\nu}$, the energy density is $\varepsilon = \langle\Gamma_{00}\rangle$ and the pressure $P = \langle\Gamma_{\mu\mu}\rangle$.

After quantizing the Dirac field and performing a detailed calculation in the RMF approximation (see Appendix 1), we obtain the energy density and pressure. The energy density is given by

$$\varepsilon = \frac{1}{2}m_\sigma^2\sigma^2 + \frac{1}{3}g_2\sigma^3 + \frac{1}{4}g_3\sigma^4 + \frac{1}{2}m_\omega^2\omega_0^2 + \frac{1}{2}m_\rho^2\rho_{03}^2$$

$$+ \frac{2}{(2\pi)^3}\sum_B \int_0^{p_{F_B}} d^3p\sqrt{p^2 + m_B^{*2}},\tag{3.44}$$

and the pressure is given by,

$$P = -\frac{1}{2}m_\sigma^2\sigma^2 - \frac{1}{3}g_2\sigma^3 - \frac{1}{4}g_3\sigma^4 + \frac{1}{2}m_\omega^2\omega_0^2 + \frac{1}{2}m_\rho^2\rho_{03}^2$$

$$+ \frac{1}{3}\frac{2}{(2\pi)^3}\sum_B \int_0^{p_{F_B}} d^3p\,\frac{p^2}{\sqrt{p^2 + m_B^{*2}}}.$$

$$\tag{3.45}$$

We consider electrons and muons in neutron star matter. Leptons are treated as noninteracting particles. The contributions of leptons in the EoS are calculated using the Lagrangian density as given by,

$$\mathcal{L}_L = \sum_{l=e^-,\mu^-} \bar{\psi}_l\left(i\gamma_\mu\partial^\mu - m_l\right)\psi_l.\tag{3.46}$$

The non-linear Walecka model of Eq. (3.33) is further extended to include the higher order self-interaction terms of ω and ρ mesons and the meson cross coupling terms as given by [150],

$$\mathcal{L}_B = \sum_{B=n,p} \bar{\Psi}_B\left(i\gamma_\mu\partial^\mu - m_B + g_{\sigma B}\sigma - g_{\omega B}\gamma_\mu\omega^\mu - \frac{1}{2}g_{\rho B}\gamma_\mu\boldsymbol{\tau}_B\cdot\boldsymbol{\rho}^\mu\right)\Psi_B$$

$$+\frac{1}{2}(\partial_\mu\sigma\partial^\mu\sigma - m_\sigma^2\sigma^2) - \frac{1}{4}\omega_{\mu\nu}\omega^{\mu\nu} + \frac{1}{2}m_\omega^2\omega_\mu\omega^\mu - \frac{1}{4}\boldsymbol{\rho}_{\mu\nu}\cdot\boldsymbol{\rho}^{\mu\nu} + \frac{1}{2}m_\rho^2\boldsymbol{\rho}_\mu\cdot\boldsymbol{\rho}^\mu - U(\sigma)$$

$$+ \frac{\kappa}{24}g_{\omega B}^4(\omega^\mu\omega_\mu)^2 + \frac{\lambda}{24}g_{\rho B}^4(\boldsymbol{\rho}^\mu\cdot\boldsymbol{\rho}_\mu)^2 + g_{\rho B}^2 f(\sigma,\omega^\mu\omega_\mu)\boldsymbol{\rho}^\mu\cdot\boldsymbol{\rho}_\mu,\tag{3.47}$$

where

$$f(\sigma,\omega^\mu\omega_\mu) = \sum_1^6 a_i\sigma^i + \sum_1^3 b_j(\omega^\mu\omega_\mu)^j.\tag{3.48}$$

The model of Eq. (3.47) has 17 parameters which provide enough freedom to fine tune the low and high density part of the isospin sector independently [151]. This Lagrangian density represents two EoS models known as the Steiner, Fischer, and Hempel SFHo with "o" standing for optimal and SFHx where "x" implies for extremal. The SFHo model fitted the most probable mass-radius curve of Ref. [152] whereas the radius of low mass neutron stars was minimized in the SFHx model resulting in a low value (23.18 MeV) of the density slope of the symmetry energy at the saturation density [151]. The deletion of the last two terms of the Lagrangian density given by Eq. (3.47) leads to the same Lagrangian density of TM1 and TMA EoS models with different parameter sets [153, 154].

3.5.1.1 Hyperon-Hyperon Interaction

The RMF model described above was extended to investigate other exotic forms of hadronic matter such as hyperon matter and antikaon condensed matter. This model can accommodate the whole baryon octet. The appearance of hyperons in the core of neutron stars is inevitable [46, 47]. The hyperon-nucleon interaction is mediated by the exchange of σ, ω and ρ mesons as given by the Lagrangian density of Eq. (3.33). As a significant population of hyperons is expected in the core, the hyperon hyperon interaction might be important. The Lagrangian density ($\mathcal{L}_{YY}$) corresponding to the hyperon-hyperon interaction is given by,

$$
\mathcal{L}_{YY} = \sum_{B=\Lambda,\Sigma,\Xi} \bar{\Psi}_B \left(g_{\sigma^*B}\sigma^* - g_{\phi B}\gamma_\mu\phi^\mu \right) \Psi_B
$$

$$
+ \frac{1}{2} \left(\partial_\mu\sigma^*\partial^\mu\sigma^* - m_{\sigma^*}^2\sigma^{*2} \right)
$$

$$
- \frac{1}{4}\phi_{\mu\nu}\phi^{\mu\nu} + \frac{1}{2}m_\phi^2\phi_\mu\phi^\mu . \tag{3.49}
$$

It is evident from double Λ hypernuclei data that the attractive $\Lambda - \Lambda$ interaction due to the exchange of σ^* meson is quite weak [155–157]. As a result, we neglect the role of σ^* meson in Eq. (3.49). However, the repulsive hyperon-hyperon interaction mediated by ϕ mesons leads to a stiffer EoS [158].

In addition to meson field equations in Eq. (3.41) modified by the presence of hyperons, we have the equation of motion for the mean field of ϕ mesons denoted by ϕ_0,

$$
m_\phi^2\phi_0 = \sum_{B=n,p,\Lambda,\Sigma^+,\Sigma^0,\Sigma^-,\Xi^0,\Xi^-} g_{\phi B}\langle\bar{\psi}_B\gamma_0\psi_B\rangle . \tag{3.50}
$$

The energy density of the hyperon matter corresponding to the Lagrangian density of Eq. (3.33) including the hyperon-hyperon interaction of Eq. (3.49) is,

$$\varepsilon = \frac{1}{2}m_\sigma^2\sigma^2 + \frac{1}{3}g_2\sigma^3 + \frac{1}{4}g_3\sigma^4 + \frac{1}{2}m_\omega^2\omega_0^2 + \frac{1}{2}m_\rho^2\rho_{03}^2 + \frac{1}{2}m_\phi^2\phi_0^2$$
$$+ \frac{2}{(2\pi)^3}\sum_{B=n,p,\Lambda,\Sigma^+,\Sigma^0,\Sigma^-,\Xi^0,\Xi^-}\int_0^{p_{F_B}} d^3p\sqrt{p^2 + m_B^{*2}}\,, \tag{3.51}$$

and the pressure is given by,

$$P = -\frac{1}{2}m_\sigma^2\sigma^2 - \frac{1}{3}g_2\sigma^3 - \frac{1}{4}g_3\sigma^4 + \frac{1}{2}m_\omega^2\omega_0^2 + \frac{1}{2}m_\rho^2\rho_{03}^2 + \frac{1}{2}m_\phi^2\phi_0^2$$
$$+ \frac{1}{3}\frac{2}{(2\pi)^3}\sum_{B=n,p,\Lambda,\Sigma^+,\Sigma^0,\Sigma^-,\Xi^0,\Xi^-}\int_0^{p_{F_B}} d^3p \frac{p^2}{\sqrt{p^2 + m_B^{*2}}}\,. \tag{3.52}$$

The constituents of matter are in chemical equilibrium due to weak interactions as well as satisfy the charge neutrality. The generalized β-decays can be written as $B_1 \longrightarrow B_2 + l + \bar{\nu}_l$ and $B_2 + l \longrightarrow B_1 + \nu_l$, where ν_l and $\bar{\nu}_l$ represent neutrino and antineutrino, respectively. The chemical equilibrium condition is expressed in terms of neutron (μ_n) and electron (μ_e) chemical potentials corresponding to the conserved baryon number and electric charge as

$$\mu_i = b_i\mu_n - q_i\mu_e\,, \tag{3.53}$$

where the i-th baryon chemical potential is explicitly given by

$$\mu_i = (k_{F_i}^2 + m_i^{*2})^{1/2} + g_{\omega i}\omega_0 + g_{\phi i}\phi_0 + g_{\rho i}\tau_{3i}\rho_{03}\,. \tag{3.54}$$

Furthermore, the charge neutrality condition is,

$$Q = \sum_B q_B n_B - n_e - n_\mu = 0\,, \tag{3.55}$$

where q_B is the baryon charge and baryon, electron, and muon number densities are denoted by n_B, n_e, and n_μ, respectively.

The knowledge of meson-baryon coupling constants in the Lagrangian densities in Eqs. (3.33) and (3.49) is needed to determine various gross properties of the dense matter. The meson-nucleon coupling constants are fixed by reproducing the nuclear matter saturation properties. We adopt different parametrizations of the RMF models as shown in Table 3.1. We represent the Walecka model with self-interaction terms of σ mesons as Glendenning (G) followed by the compression modulus and two values of effective masses 0.78 and 0.7 denoted by a and b, respectively. Empirical values of nuclear matter properties are recorded in the last row of the table.

The vector coupling constants for hyperons are determined from the SU(6) symmetry as [61, 159],

$$\frac{1}{2}g_{\omega\Lambda} = \frac{1}{2}g_{\omega\Sigma} = g_{\omega\Xi} = \frac{1}{3}g_{\omega N},$$

$$\frac{1}{2}g_{\rho\Sigma} = g_{\rho\Xi} = g_{\rho N}; \quad g_{\rho\Lambda} = 0,$$

$$2g_{\phi\Lambda} = 2g_{\phi\Sigma} = g_{\phi\Xi} = -\frac{2\sqrt{2}}{3}g_{\omega N}. \tag{3.56}$$

Nucleons do not couple to ϕ mesons, i.e. $g_{\phi N} = 0$.

We estimate the scalar meson (σ) coupling to hyperons from the potential depth of a hyperon (Y) in the saturated nuclear matter as

$$U_Y^N(n_0) = -g_{\sigma Y}\sigma + g_{\omega Y}\omega_0. \tag{3.57}$$

The analysis of Λ-hypernuclei data leads to a well depth of Λ in the normal nuclear matter $U_\Lambda^N(n_0) = -30\,\text{MeV}$ [160, 161]. Furthermore, the analysis of a few Ξ-hypernuclei events in emulsion experiments predicted a Ξ well depth of $U_\Xi^N(n_0) = -18\,\text{MeV}$ [162–165] in the normal nuclear matter. However, the situation for the Σ potential depth in the normal nuclear matter is not quite clear because Σ-hypernuclei data are scarce. The only known bound Σ-hypernucleus is $_\Sigma^4 He$ [166]. The analysis of Σ^- atomic data resulted in a strong isoscalar repulsion in Σ-nuclear matter interaction [167]. This kind of repulsive interaction might push their appearance to higher densities or rule out the onset of Σ hyperons in dense matter.

3.5.2 Bose–Einstein Condensates of (Anti)kaons

Kaplan and Nelson explored using a chiral $SU(3)_L \times SU(3)_R$ Lagrangian that K^- meson may undergo the Bose–Einstein condensation in dense hadronic matter formed in heavy ion collisions due to the attractive s-wave kaon-nucleon interaction [51, 52]. The strongly attractive K^--nucleon interaction could lower the effective mass of K^- mesons in dense matter resulting in the decrease of the in-medium energy of K^- mesons. The s-wave K^- condensation sets in when the in-medium energy of K^- mesons equals its chemical potential. Later, others studied the K^- condensation in the core of a protoneutron star using the chiral model as well as the traditional meson exchange models [53–64]. The onset of antikaon condensation in neutron star matter is sensitive to the nuclear equation of state and also on the depth of the attractive antikaon optical potential. Banik and Bandyopadhyay investigated both K^- and $\bar{K}^0$ in neutron star matter for the first time [64]. It is worth mentioning here that the s-wave pion-nucleon interaction is repulsive whereas the p-wave pion-nucleon interaction is attractive. It is expected that the antikaon condensation is energetically favored. We describe the antikaon condensation using the meson exchange model in the following paragraphs.

The pure antikaon condensed matter is made of baryons, antikaons, and leptons. This is in chemical equilibrium under weak interactions and charge neutral. In this matter, baryons are embedded in the condensate and behave differently than the baryons in the pure hadronic phase without the condensate [54, 64]. The effective mass of a baryon in the condensed phase is smaller than that of the latter case. The effective baryon mass decreases in both phases. The baryon-baryon interaction is described by the Lagrangian density of Eq. (3.33). On other hand, baryon-(anti)kaon interaction can be treated in the same footing as the baryon-baryon interaction. The Lagrangian density of (anti)kaons is written in the minimal coupling prescription as [54, 62, 64, 168, 169],

$$\mathcal{L}_K = D_\mu^* \bar{K} D^\mu K - m_K^{*2} \bar{K} K \,, \tag{3.58}$$

where K and $\bar{K}$ denote kaon and (anti)kaon doublets; the covariant derivative is $D_\mu = \partial_\mu + i g_{\omega K} \omega_\mu + i g_{\rho K} \boldsymbol{\tau}_K \cdot \boldsymbol{\rho}_\mu + i g_{\phi K} \phi_\mu$ and the effective mass of antikaons is $m_K^* = m_K - g_{\sigma K} \sigma$.

Mean meson field equations of motion in the antikaon condensate phase are

$$m_\sigma^2 \sigma = -\frac{\partial U}{\partial \sigma} + \sum_B g_{\sigma B} n_B^{\bar{K},S} + g_{\sigma K} \sum_{\bar{K}} n_{\bar{K}} \,, \tag{3.59}$$

$$m_\omega^2 \omega_0 = \sum_B g_{\omega B} n_B^{\bar{K}} - g_{\omega K} \sum_{\bar{K}} n_{\bar{K}} \,, \tag{3.60}$$

$$m_\rho^2 \rho_{03} = \sum_B g_{\rho B} \tau_{3B} n_B^{\bar{K}} + g_{\rho K} \sum_{\bar{K}} \tau_{3\bar{K}} n_{\bar{K}} \,, \tag{3.61}$$

$$m_\phi^2 \phi_0 = \sum_B g_{\phi B} n_B^{\bar{K}} - g_{\phi K} \sum_{\bar{K}} n_{\bar{K}} \,, \tag{3.62}$$

where the isospin projection $\tau_{3\bar{K}} = \mp 1$ for K^- ($-$ sign) and $\bar{K}^0$ ($+$ sign) mesons.

Here the scalar and number density of baryon B in the antikaon condensed phase are, respectively,

$$n_B^{\bar{K},S} = \frac{1}{\pi^2} \int_0^{k_{F_B}} \frac{m_B^*}{(k^2 + m_B^{*2})^{1/2}} k^2 \, dk \,, \tag{3.63}$$

$$n_B^{\bar{K}} = \frac{k_{F_B}^3}{3\pi^2} \,, \tag{3.64}$$

whereas the scalar and vector densities of antikaons are same as

$$n_{K^-,\bar{K}^0} = 2\left(\omega_{K^-,\bar{K}^0} + g_{\omega K}\omega_0 + g_{\phi K}\phi_0 \mp \tau_{3\bar{K}}g_{\rho K}\rho_{03}\right)\bar{K}K = 2m_K^*\bar{K}K \ . \tag{3.65}$$

The in-medium energies of $\bar{K} \equiv (K^-, \bar{K}^0)$ for s-wave ($\mathbf{k} = 0$) condensation are given by

$$\omega_{K^-,\bar{K}^0} = m_K^* - g_{\omega K}\omega_0 - g_{\phi K}\phi_0 \pm \tau_{3\bar{K}}g_{\rho K}\rho_{03} \ . \tag{3.66}$$

The total energy density in the antikaon condensed phase taking into account the contributions of baryons, leptons, and antikaons is,

$$\varepsilon^{\bar{K}} = \frac{1}{2}m_\sigma^2\sigma^2 + \frac{1}{3}g_2\sigma^3 + \frac{1}{4}g_3\sigma^4 + \frac{1}{2}m_\omega^2\omega_0^2 + \frac{1}{2}m_\phi^2\phi_0^2 + \frac{1}{2}m_\rho^2\rho_{03}^2$$

$$+ \sum_B \frac{1}{\pi^2}\int_0^{k_{F_B}}(k^2 + m_B^{*2})^{1/2}k^2\,dk + \sum_l \frac{1}{\pi^2}\int_0^{K_{F_l}}(k^2 + m_l^2)^{1/2}k^2\,dk$$

$$+ m_K^*\left(n_{K^-} + n_{\bar{K}^0}\right) \ . \tag{3.67}$$

As the antikaon condensate is a s-wave condensate, the pressure does not contain any contribution of the antikaon condensate explicitly as shown below,

$$P^{\bar{K}} = -\frac{1}{2}m_\sigma^2\sigma^2 - \frac{1}{3}g_2\sigma^3 - \frac{1}{4}g_3\sigma^4 + \frac{1}{2}m_\omega^2\omega_0^2 + \frac{1}{2}m_\phi^2\phi_0^2 + \frac{1}{2}m_\rho^2\rho_{03}^2$$

$$+ \frac{1}{3}\sum_B \frac{1}{\pi^2}\int_0^{k_{F_B}}\frac{k^4\,dk}{(k^2 + m_B^{*2})^{1/2}} + \frac{1}{3}\sum_l \frac{1}{\pi^2}\int_0^{K_{F_l}}\frac{k^4\,dk}{(k^2 + m_l^2)^{1/2}} \ . \tag{3.68}$$

The charge neutrality in the antikaon condensed matter is

$$Q^{\bar{K}} = \sum_B q_B n_B^{\bar{K}} - n_{\bar{K}}^- - n_e - n_\mu = 0. \tag{3.69}$$

The β-equilibrium due to strangeness changing processes in the medium may occur, for example, $N \rightleftharpoons N + \bar{K}$ and $e^- \rightleftharpoons K^- + \nu_e$, where $N \equiv (n, p)$ and $\bar{K} \equiv (K^-, \bar{K}^0)$ denote the isospin doublets for nucleons and antikaons, respectively. The chemical equilibrium can be written as

$$\mu_n - \mu_p = \mu_{K^-} = \mu_e \ ,$$

$$\mu_{\bar{K}^0} = 0 \ , \tag{3.70}$$

where μ_{K^-} and $\mu_{\bar{K}^0}$ are the chemical potentials of K^- and $\bar{K}^0$ mesons, respectively. The onset of K^- condensation happens when the in-medium energy of K^- mesons

(ω_{K^-}) equals to its chemical potential (μ_{K^-}) which, is again equal to μ_e. The threshold condition of $\bar{K}^0$ condensation is given by its in-medium energy satisfying the condition $\omega_{\bar{K}^0} = \mu_{\bar{K}^0} = 0$.

The coupling constants of meson fields with (anti)kaons are estimated theoretically as well as phenomenologically. The vector coupling constants of Eq. (3.58) are determined following the quark model and isospin counting rule,

$$g_{\omega K} = \frac{1}{3}g_{\omega N} \quad \text{and} \quad g_{\rho K} = g_{\rho N} . \tag{3.71}$$

The scalar coupling constant is extracted from the real part of the K^- optical potential depth in the normal nuclear matter density

$$U_{\bar{K}}(n_0) = -g_{\sigma K}\sigma - g_{\omega K}\omega_0 . \tag{3.72}$$

As the strange ϕ meson field couples with (anti)kaons, the vector ϕ meson coupling with (anti)kaons is obtained from the SU(3) relation as $\sqrt{2}g_{\phi K} = g_{\pi\pi\rho} = 6.04$ [61]. It was noted that antikaons experienced an attractive potential and kaons had a repulsive interaction in nuclear matter [167, 170–175]. The real part of the antikaon optical potential as large as $U_{\bar{K}} = -180 \pm 20$ MeV at the normal nuclear matter density was obtained from the analysis of K^- atomic data in the hybrid model; but it was repulsive at low density following the low density theorem [172]. The antikaon potential depth in the coupled channel calculation was $U_{\bar{K}} = -100$ MeV whereas the chirally motivated coupled channel approach yielded the potential depth of $U_{\bar{K}} = -120$ MeV [173, 174]. It was argued that the different treatments of $\Lambda(1405)$ resonance which is considered to be an unstable $\bar{K}N$ bound state just below the K^-p threshold, were responsible for the wide range of antikaon optical potential depth values in various calculations.

Antikaon condensation could be a first or second order phase transition. Here we give an example where the K^- condensation was treated as a first order phase transition followed by the second order $\bar{K}^0$ condensation [64]. The first order phase transition was governed by the Gibbs phase rules [54]. The onsets of K^- and $\bar{K}^0$ condensates depend sensitively on the EoS. A stiffer hadronic EoS leads to the early onset of an antikaon condensate. Furthermore, the early appearance of an antikaon condensate might delay the population of hyperons or vice versa. Particle abundances are shown in Fig. 3.4 for the Glendenning and Moszkowski parametrization of the RMF model known as the GM1 set [176]. This calculation was performed for the antikaon optical potential depth of -160 MeV at the normal nuclear matter density. The mixed phase of the pure hadronic matter described by Eq. (3.33) and the K^- condensed matter appears at $2.23 \, n_0$. It is noted from Fig. 3.4 that Λ hyperons appear first inside the mixed phase at $2.51 \, n_0$ when the threshold condition $\mu_\Lambda = \mu_n$ is satisfied. The K^- meson number density in the condensate grows rapidly replacing negatively charged leptons, i.e. electrons and muons. The $\bar{K}^0$ condensate is populated at $4.1 \, n_0$ just after the mixed phase ends at $4.0 \, n_0$. Neutron and proton abundances become equal immediately after the appearance

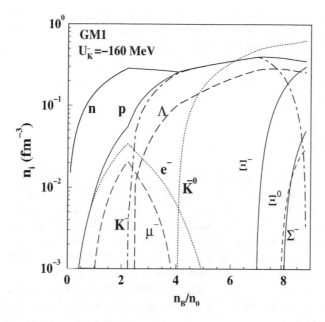

Fig. 3.4 Number densities of different particle species in the β-equilibrated matter including hyperons along with K^- and K^0 condensates are shown as a function of total baryon density for the GM1 model and the antikaon optical potential depth of -160 MeV. Reprinted figure with permission from Banik and Bandyopadhyay [64] ©2018 by the American Physical Society (APS). https://doi.org/10.1103/PhysRevC.64.055805

of the $\bar{K}^0$ condensate. Heavier hyperons like Ξ^-, Ξ^0, Σ^- are populated at higher densities in the pure condensed phase. With the appearance of negatively charged hyperons, the number density of K^- mesons in the condensate drops. This might happen beyond the central density of a neutron star.

3.5.3 Quark Matter

The possibility of the existence of quark matter inside the neutron star was first proposed by N. Itoh [65]. It is expected that hadrons would dissolve into their fundamental constituents at baryon densities of a few times the saturation density in the core of neutron stars. Further Edward Witten conjectured that the strange quark matter made of up, down, and strange quarks in equal proportions might be the true ground state of the strongly interacting matter in the quantum chromodynamics (QCD) [68]. In this case, the energy per baryon of the strange quark matter is less than the energy per nucleon of the stable Fe nucleus [67, 68].

The quark matter can be studied using the perturbative QCD at very high baryon densities when the asymptotic freedom is reached [69]. However, the baryon density in the central part of neutron stars might not be enough to attain the asymptotic freedom when quarks can be treated as weakly interacting. The quark confinement at lower energies or baryon densities in the neutron star interior is an issue. This non-perturbative effect of the QCD is dealt with in the Lattice Gauge Theory. The Lattice QCD for finite baryon density is a difficult problem. We discuss below various phenomenological models of the quark matter.

3.5.3.1 Unpaired Quark Matter

The most popular and widely used model for the calculation of quark matter is the Massachusetts Institute of Technology (MIT) Bag model [177]. The bag provides the necessary confinement for quarks inside hadrons. Quarks are treated as Fermi gas inside the bag. The QCD vacuum inside the volume where quarks reside is expelled at the cost of energy. This energy per unit volume is the bag constant (B). The energy density of the quark matter in the bag model consists of kinetic energies of quarks and the bag constant which is treated as a parameter.

The EoS of unpaired quark matter is obtained from the Lagrangian density of noninteracting quarks in the bag at zero temperature. The Lagrangian density of free quarks is given by,

$$\mathcal{L}_q = \sum_{f=u,d,s} \bar{\psi}_f \left(i\gamma_\mu \partial^\mu - m_f \right) \psi_f , \qquad (3.73)$$

where f stands for quark flavor and m_f is the mass of a quark flavor. The kinetic energy of a quark is

$$E_f = \sqrt{(p^2 + m_f^2)} . \qquad (3.74)$$

The baryon density, energy density, and pressure of the quark matter are calculated in the following way. The baryon density

$$n_B^q = \sum_f \frac{1}{3} \frac{\gamma_f}{2\pi^2} \int_0^{p_{F_f}} p^2 \, dp = \frac{1}{3\pi^2} \sum_f p_{F_f}^3 , \qquad (3.75)$$

where the degeneracy of each quark flavor with two spins and three colors is $\gamma_f = 2 \times 3$, the factor $1/3$ is the baryon number of each quark and p_{F_f} is the Fermi momentum of a quark flavor.

The energy density

$$\varepsilon = B + \sum_f \frac{\gamma_f}{2\pi^2} \int_0^{p_{F_f}} E_f p^2 \, dp , \qquad (3.76)$$

simplifies to

$$\varepsilon = B + \sum_f \frac{3}{4\pi^2} \left[\mu_f p_{F_f} \left(\mu_f^2 - \frac{1}{2} m_f^2 \right) + \frac{1}{2} m_f^4 \ln \left(\frac{\mu_f + p_{F_f}}{m_f} \right) \right], \quad (3.77)$$

where the quark chemical potential is $\mu_f = \sqrt{p_{F_f}^2 + m_f^2}$.

The pressure

$$P = -B + \sum_f \frac{1}{3} \frac{\gamma_f}{2\pi^2} \int_0^{p_{F_f}} \frac{p^4}{E_f} dp, \quad (3.78)$$

after simplification becomes,

$$P = -B + \sum_f \frac{1}{4\pi^2} \left[\mu_f p_{F_f} \left(\mu_f^2 - \frac{5}{2} m_f^2 \right) + \frac{3}{2} m_f^4 \ln \left(\frac{\mu_f + p_{F_f}}{m_f} \right) \right]. \quad (3.79)$$

The following weak interaction processes occur in the quark matter,

$$d \rightarrow u + e + \bar{\nu}_e; \quad s \rightarrow u + e^- + \bar{\nu}_e. \quad (3.80)$$

These lead to the chemical equilibrium conditions, $\mu_d = \mu_u + \mu_e$ and $\mu_s = \mu_d$. Neutrinos freely come out of the star. The charge neutrality is given by,

$$\frac{2}{3} n_u - \frac{1}{3} n_d - \frac{1}{3} n_s - n_e = 0, \quad (3.81)$$

where each factor associated with quark number densities implies the electric charge of the corresponding quark flavor and n_e is the electron number density.

The perturbative QCD corrections to the quark matter EoS due to the quark-quark interaction mediated by gluons was first estimated to $O(\alpha_s^2)$ by Freedman and McLerran and later followed by others [67, 89, 90]. Fraga et al. also obtained the pressure of the quark matter with three massless quarks up to α_s^2 in a bag inspired model,

$$P = -\Omega_{QM} = \frac{3}{4\pi^2} a_4 \mu^4 - B_{eff}, \quad (3.82)$$

where α_s is the QCD coupling constant [91]. They estimated the effective bag constant (B_{eff}) matching Eq. (3.82) with their perturbative calculations to $O(\alpha_s^2)$ for the quark chemical potential $\mu \sim 300$–600 MeV. Although B_{eff} varies with the renormalization scale in the range 145–233 MeV, $a_4 \sim 0.626$–0.628 remains constant. Later Alford et al. proposed a simple parametrization of the quark matter with the grand thermodynamic potential as given by [178],

$$\Omega_{QM} = -\frac{3}{4\pi^2}a_4\mu^4 + \frac{3}{4\pi^2}a_2\mu^2 - B_{eff} , \tag{3.83}$$

where $a_4 = 1 - c$ and c is the perturbative QCD correction. For the massless noninteracting quark matter, $c = 0$, i.e. $a_4 = 1$. Here a_2 is important when non-zero strange quark mass and quark-quark pairing are taken into account. In case of the unpaired quark matter, $a_2^{1/2}$ is the strange quark mass [178]. It is noted that a_4 plays an important role in determining the maximum masses of hybrid stars which include the quark matter in the core. For $a_4 \sim 0.65$, the hybrid stars of 2 $M_\odot$ can be obtained [178, 179].

3.5.3.2 Color Superconductivity in Quark Matter

The quark matter may be a color superconductor as quarks near their Fermi surfaces form Cooper pairs because of the attractive quark-quark interaction in the antisymmetric color channel [180–186]. The color gauge symmetry is broken due to the formation of diquark condensates. At very high densities, massless quarks with all three flavors and colors would pair up to produce an energetically favored state called the color-flavor-locked (CFL) phase[69]. In case of a large strange quark mass, a less symmetric 2-flavor superconducting (2SC) phase appears with the formation of pairs between light u and d quarks.

As pairs of quarks cannot be color and flavor singlets, the formation of diquark condensates breaks the color and flavor $SU(3)_{color} \times SU(3)_L \times SU(3)_R$ symmetries to $SU(3)_{color+L+R}$. Massless Nambu-Goldstone bosons are generated due to the chiral symmetry breaking by the CFL condensate [69, 187, 188]. Further, all the quark modes acquire gaps at their Fermi surfaces. An additional scalar Nambu-Goldstone boson is associated with the spontaneous breakdown of the baryon number symmetry. The symmetric CFL phase under stresses such as non-zero strange quark mass and electron chemical potential was also investigated [187, 188]. The CFL phase might relax under those stresses giving rise to the meson condensation, for example, K^0 condensation [187].

Let us consider the pure CFL quark matter consisting of paired quarks of all flavors and colors. The grand thermodynamic potential for electric and color charge neutral CFL quark matter to order Δ^2 is given by [69, 189, 190]

$$\Omega_{QM}^{CFL} = \frac{6}{\pi^2}\int_0^\nu p^2(p-\mu)dp + \frac{3}{\pi^2}\int_0^\nu p^2\left(\sqrt{p^2+m_s^2}-\mu\right)dp - \frac{3\Delta^2\mu^2}{\pi^2} + B , \tag{3.84}$$

where Δ denotes the color superconducting gap and B is the bag constant. The first two terms of Eq. (3.84) are contributions to the thermodynamic potential of the unpaired quark matter. All quarks that participate in the pairing have a common

Fermi momentum (ν) to minimize the thermodynamic potential of the unpaired quark matter [69, 189, 190]. The contribution of the CFL condensate to Ω_{QM}^{CFL} is given by the third term. The common Fermi momentum is defined by

$$\nu = 2\mu - \sqrt{\mu^2 + \frac{m_s^2}{3}}, \tag{3.85}$$

where the average quark chemical potential is μ and the strange quark mass is m_s. It follows from the pairing ansatz in the CFL phase that

$$n_u = n_r, \qquad n_d = n_g \qquad \text{and} \qquad n_s = n_b, \tag{3.86}$$

where n_r, n_g, n_b and n_u, n_d, n_s are color and flavor number densities, respectively [186]. The quark number densities in the CFL quark matter are $n_u = n_d = n_s = (\nu^3 + 2\Delta^2\mu)/\pi^2$. The pairing gap Δ has a value in the range 10–100 MeV [69]. The relation in Eq. (3.86) demonstrates that the color neutrality automatically enforces the electric charge neutrality in the CFL phase. Consequently the electric charge chemical potential is $\mu_e = 0$ and the color chemical potential of color charge T_3 is $\mu_3 = 0$. On the other hand, the color chemical potential (μ_8) of color charge T_8 has a non-zero value to maintain the color charge neutrality in the CFL phase [185, 186]. The pressure in the CFL quark matter is $P_{QM}^{CFL} = -\Omega_{CFL}^{QM}$. The energy density in the CFL phase is obtained from the Gibbs–Duhem relation $\varepsilon_{QM}^{CFL} = \sum_f n_f \mu_f - P_{QM}^{CFL}$.

3.5.3.3 Nambu-Jona-Lasinio Model for Quark Matter

The SU(3) Nambu-Jona-Lasino (NJL) model is being widely used for the study of quark matter. The NJL model captures several basic aspects of the QCD [191, 192]. The NJL model, being an effective theory of the QCD, describes the interactions between quarks. It is an efficient tool to study the dynamical mass generation, the formation of quark pair condensates through the spontaneous breaking of the chiral symmetry. Here we introduce the local SU(3) NJL model. The grand thermodynamic potential in the mean field approximation at zero temperature is obtained from the effective interaction of the local three-flavor NJL model with vector interactions and is given by [193, 194]

$$\Omega_{QM}^{NJL}(M_f, \mu) = G_S \sum_{f=u,d,s} \langle \bar{\psi}_f \psi_f \rangle^2 + 4H \langle \bar{\psi}_u \psi_u \rangle \langle \bar{\psi}_d \psi_d \rangle \langle \bar{\psi}_s \psi_s \rangle$$

$$-2N_c \sum_{f=u,d,s} \int_\Lambda \frac{d^3 p}{(2\pi)^3} E_f$$

$$-\frac{N_c}{3\pi^2} \sum_{f=u,d,s} \int_0^{p_{F_f}} dp \frac{p^4}{E_f} - G_V \sum_f \rho_f^2, \tag{3.87}$$

where number of quark colors $N_c = 3$, $E_f = \sqrt{\mathbf{p}^2 + M_f^2}$, and p_{F_f} is the Fermi momentum, G and H are coupling constants and Λ is the three momentum ultraviolet cutoff parameter. The constituent quark mass of each flavor (M_f) is defined as,

$$M_f = m_f - 2G_S \langle \bar{\psi}_f \psi_f \rangle - 2H \langle \bar{\psi}_j \psi_j \rangle \langle \bar{\psi}_k \psi_k \rangle , \tag{3.88}$$

where f, j, k implying cyclic permutations. The quark chemical potential

$$\mu_f = E_F - 2G_V \rho_f , \tag{3.89}$$

where E_F is the Fermi energy. The quark number density of each flavor is,

$$\rho_f = \frac{N_c}{3\pi^2}[(E_F - 2G_V \rho_f)^2 - M_f^2]^{3/2} . \tag{3.90}$$

The thermodynamic potential is minimized,

$$\frac{\partial \Omega_{QM}^{NJL}}{\partial \langle \bar{\psi}_f \psi_f \rangle} = 0 , \quad f = u, d, s , \tag{3.91}$$

to obtain the quark condensates $\langle \bar{\psi}_f \psi_f \rangle$. The nonlocal extension of the SU(3) NJL model is also adopted for the calculation of the quark matter [193, 194].

3.5.4 Density Dependent Hadronic Field Theory at Finite Temperature

Now we apply the density dependent relativistic hadronic (DDRH) field theory to describe the dense matter in the neutron star core at finite temperature. The β-equilibrated and charge neutral hadronic matter described within the framework of the DDRH model is made of all species of the baryon octet, electrons and muons. Therefore, the total Lagrangian density in the hadronic phase is written as $\mathcal{L} = \mathcal{L}_B + \mathcal{L}_L$. In the DDRH model, the baryon-baryon interaction is given by the Lagrangian density ($\mathcal{L}_B$) [158, 168, 195],

$$\mathcal{L}_B = \sum_{F=N,\Lambda,\Sigma,\Xi} \bar{\Psi}_F \left(i\gamma_\mu \partial^\mu - m_F + g_{\sigma F}\sigma - g_{\omega F}\gamma_\mu \omega^\mu \right.$$
$$\left. - g_{\rho F}\gamma_\mu \boldsymbol{\tau}_F \cdot \boldsymbol{\rho}^\mu - g_{\phi F}\gamma_\mu \phi^\mu \right) \Psi_F$$
$$+ \frac{1}{2} \left(\partial_\mu \sigma \partial^\mu \sigma - m_\sigma^2 \sigma^2 \right) - \frac{1}{4}\omega_{\mu\nu}\omega^{\mu\nu} + \frac{1}{2}m_\omega^2 \omega_\mu \omega^\mu$$

$$-\frac{1}{4}\rho_{\mu\nu}\cdot\rho^{\mu\nu}+\frac{1}{2}m_\rho^2\rho_\mu\cdot\rho^\mu$$

$$-\frac{1}{4}\phi_{\mu\nu}\phi^{\mu\nu}+\frac{1}{2}m_\phi^2\phi_\mu\phi^\mu\,, \tag{3.92}$$

and the lepton part is given by Eq. (3.46). The field strength tensors for vector mesons are given by Eq. (3.34).

As in Sect. 3.5.1 for Ψ_B, Ψ_F denotes the isospin multiplets for baryons and the sum goes over baryon multiplets $F = N, \Lambda, \Sigma, \Xi$; m_B is the bare baryon mass and τ_F is the isospin operator. In addition to σ, ω, and ρ mesons, the strange ϕ mesons are also included for the hyperon-hyperon interaction. The structure of the DDRH Lagrangian density is similar to that of the RMF model already discussed in Sect. 3.5.1. However, there are important differences between those models. In the RMF calculations with density independent meson-baryon coupling constants, non-linear self-interaction terms of scalar and vector fields are considered for higher order density dependent contributions. But this is not the case when meson-baryon vertices $g_{\alpha F}$ with α denoting σ, ω, ρ, and ϕ fields, are dependent on Lorentz scalar functionals of baryon field operators. There are two choices for the density dependence of meson-baryon couplings—the scalar density dependence (SDD) and the vector density dependence (VDD) [124]. Here we consider meson-baryon couplings $g_{\alpha B}(\hat{n})$ to depend on the vector density because it gives a more natural connection to the parametrization of DB vertices. For the VDD case, the density operator $\hat{n}$ is given by, $\hat{n} = \sqrt{\hat{j}_\mu \hat{j}^\mu}$, where $\hat{j}_\mu = \bar{\Psi}\gamma_\mu\Psi$.

The grand canonical partition function in the mean field approximation can be written as [196],

$$\ln Z_B = \beta V\left[-\frac{1}{2}m_\sigma^2\sigma^2+\frac{1}{2}m_\omega^2\omega_0^2+\frac{1}{2}m_\rho^2\rho_{03}^2+\frac{1}{2}m_\phi^2\phi_0^2\right.$$

$$\left.+\Sigma^r\sum_{i=n,p,\Lambda,\Sigma^+,\Sigma^0,\Sigma^-,\Xi^0,\Xi^-}n_i\right]$$

$$+2V\sum_{i=n,p,\Lambda,\Sigma^+,\Sigma^0,\Sigma^-,\Xi^0,\Xi^-}\int\frac{d^3k}{(2\pi)^3}[\ln(1+e^{-\beta(E_i^*-\nu_i)})$$

$$+\ln(1+e^{-\beta(E_i^*+\nu_i)})]\,, \tag{3.93}$$

where the temperature is $\beta = 1/T$, $E_i^* = \sqrt{(k^2 + m_i^{*2})}$ and the effective mass of i-th baryon $m_i^* = m_i - g_{\sigma i}\sigma$. The total grand canonical partition function of the system is $Z = Z_B Z_L$ where Z_L denotes the grand canonical partition function for noninteracting leptons. The equations of motion for meson and baryon fields are obtained by extremizing Z_B.

We carry out this calculation in the mean field approximation as it is done in Sect. 3.5.1. In this approximation, vertex functionals take simpler forms using Wick's theorem [124, 197]. The operator $\hat{n}$ is replaced by the ground state expectation value of $\hat{n}$, i.e. $\langle \hat{n} \rangle = n$. Hence meson-baryon vertices become function of total baryon density

$$\langle g_{\alpha F}(\hat{n}) \rangle = g_{\alpha F}(\langle \hat{n} \rangle) = g_{\alpha F}(n). \tag{3.94}$$

This is the vector density dependence of vertices [109, 124, 126]. As the meson-baryon couplings $g_{\alpha F}$s are density dependent, the variation of $\mathcal{L}$ with respect to $\bar{\Psi}_F$ gives an additional term,

$$\frac{\delta \mathcal{L}_B}{\delta \bar{\Psi}_F} = \frac{\partial \mathcal{L}_B}{\partial \bar{\Psi}_F} + \frac{\partial \mathcal{L}_B}{\partial n_i} \frac{\delta n_i}{\delta \bar{\Psi}_F}. \tag{3.95}$$

The rearrangement term $\Sigma^r = \sum_F \frac{\partial \mathcal{L}_B}{\partial n_i} \frac{\delta n_i}{\delta \bar{\Psi}_F}$ which originates from the second term of Eq. (3.95), naturally introduces an additional contribution to the vector self-energy [109, 110, 124, 126]. This is an important difference between the RMF and DDRH theory. The rearrangement term which takes care of many-body correlations has the form,

$$\Sigma^r = \sum_{i=n,p,\Lambda,\Sigma^+,\Sigma^0,\Sigma^-,\Xi^0,\Xi^-} \left[-\frac{\partial g_{\sigma i}}{\partial n_i} \sigma n_i^s + \frac{\partial g_{\omega i}}{\partial n_i} \omega_0 n_i + \frac{\partial g_{\rho i}}{\partial n_i} \tau_{3i} \rho_{03} n_i \right.$$
$$\left. + \frac{\partial g_{\phi B}}{\partial n_i} \phi_0 n_i \right]. \tag{3.96}$$

The chemical potential of i-th baryon is given by

$$\mu_i = \nu_i + g_{\omega i} \omega_0 + g_{\rho i} \tau_{3i} \rho_{03} + g_{\phi i} \phi_0 + \Sigma^r. \tag{3.97}$$

Meson field equations of motion are given by

$$m_\sigma^2 \sigma = \sum_i g_{\sigma i} n_i^s, \tag{3.98}$$

$$m_\omega^2 \omega_0 = \sum_i g_{\omega i} n_i, \tag{3.99}$$

$$m_\rho^2 \rho_{03} = \sum_i g_{\rho i} \tau_{3i} n_i, \tag{3.100}$$

$$m_\phi^2 \phi_0 = \sum_i g_{\phi i} n_i, \tag{3.101}$$

where τ_{3i} is the isospin projection of i-th baryon The number density of i-th baryon is

$$n_i = 2 \int \frac{d^3k}{(2\pi)^3} \left(\frac{1}{e^{\beta(E_i^* - \nu_i)} + 1} - \frac{1}{e^{\beta(E_i^* + \nu_i)} + 1} \right). \qquad (3.102)$$

The scalar density (n_i^s) of i-th baryon is

$$n_i^s = 2 \int \frac{d^3k}{(2\pi)^3} \frac{m_B^*}{E_i^*} \left(\frac{1}{e^{\beta(E_i^* - \nu_i)} + 1} + \frac{1}{e^{\beta(E_i^* + \nu_i)} + 1} \right). \qquad (3.103)$$

The rearrangement contribution modifies the baryon field equation compared to the RMF case [106],

$$[\gamma_\mu \left(i \partial^\mu - g_{\omega i} \omega_0 - g_{\rho i} \tau_{3i} \rho_{03} - g_{\phi i} \phi_0 - \Sigma^r \right) - (m_i - g_{\sigma i} \sigma)] \psi_i = 0. \qquad (3.104)$$

Here ψ_i is the Dirac spinor for the i-th baryon, and the effective baryon mass is $m_i^* = m_i - g_{\sigma i} \sigma$.

The energy density of baryons is,

$$\epsilon = \frac{1}{2} m_\sigma^2 \sigma^2 + \frac{1}{2} m_\omega^2 \omega_0^2 + \frac{1}{2} m_\rho^2 \rho_{03}^2 + \frac{1}{2} m_\phi^2 \phi_0^2$$

$$+ 2 \sum_i \int \frac{d^3k}{(2\pi)^3} E_i^* \left(\frac{1}{e^{\beta(E_i^* - \nu_i)} + 1} + \frac{1}{e^{\beta(E_i^* + \nu_i)} + 1} \right), \qquad (3.105)$$

and the baryon pressure is written as $P = TV^{-1} \ln Z_B$. The entropy density of baryons follows from the relation

$$S = \beta \left(\epsilon + P - \sum_i \mu_i n_i \right), \qquad (3.106)$$

and the entropy per baryon is $s = S/n_B$ where $n_B = \sum_i n_i$ is the total baryon density.

Similarly we calculate number densities, energy densities and pressures of noninteracting leptons and their antiparticles using the following partition function,

$$\ln Z_L = 2V \sum_l \int \frac{d^3k}{(2\pi)^3} [\ln(1 + e^{-\beta(E_l - \mu_l)}) + \ln(1 + e^{-\beta(E_l + \mu_l)})], \qquad (3.107)$$

where E_l and μ_l are energy and chemical potential of leptons.

Meson-nucleon couplings in the DDRH model are density dependent. Here density dependent meson-nucleon couplings adopted to describe the nuclear matter properties is denoted as the DD2 parameter set [122, 195]. The density dependent couplings $g_{\sigma N}$ and $g\omega N$ have the following forms,

$$g_{\alpha N} = g_{\alpha N}(n_0) f_\alpha(x) ,$$

$$f_\alpha(n_B/n_0) = a_\alpha \frac{1 + b_\alpha(x + d_\alpha)^2}{1 + c_\alpha(x + d_\alpha)^2} , \qquad (3.108)$$

where n_0 is the saturation density, $\alpha = \sigma, \omega$ and $x = n_B/n_0$. In case of ρ mesons, it is given by,

$$g_{\rho N} = g_{\rho N}(n_0) \exp[-a_\rho(x - 1)] . \qquad (3.109)$$

Next we obtain the scalar meson coupling to hyperons (Y) $(g_{\sigma Y})$ from the potential depth of a hyperon in the normal nuclear matter. In the density dependent RMF model, the potential depth of a hyperon in the saturated nuclear matter is given by

$$U_Y^N(n_0) = g_{\omega Y}\omega_0 + \Sigma_N^r - g_{\sigma Y}\sigma , \qquad (3.110)$$

where Σ_N^r is the contribution of only nucleons in the rearrangement term as given by Eq. (3.96). The vector meson-hyperon couplings are given by the SU(6) symmetry as discussed already.

Coefficients in Eqs. (3.108) and (3.109), the a_αs, the b_αs, the c_αs, and the d_αs as well as the a_ρ, the saturation density, meson-nucleon couplings at the saturation density, the mass of σ mesons are extracted by fitting the properties of finite nuclei [195]. The symmetric nuclear matter properties at the saturation density ($n_0 = 0.149065\,\mathrm{fm}^{-3}$) are consistent with the experimental values [70]. The symmetry energy (32.73 MeV) and its density slope (57.94 MeV) are in good agreement with experimental findings and observations of neutron stars [198–200]. Furthermore, the DD2 EoS agrees reasonably well with that of the pure neutron matter in the chiral effective field theory [70, 201]. The Lagrangian density of Eq. (3.92) including Λ hyperons and ϕ mesons leads to the widely used EoS of Banik, Hempel and Bandyopadhyay (BHB$\Lambda\phi$) [158].

3.5.5 Antikaon Condensation at Finite Temperature

We extend the zero temperature calculation of antikaon condensation to the case of finite temperature as discussed in Sect. 3.5.2. The Lagrangian density of Eq. (3.58) is the starting point in this case.

The partition function for (anti)kaons is given by [31]

$$\ln Z_K = 2V \int \frac{d^3k}{(2\pi)^3} [\ln(1 + e^{-\beta(\omega_{K^-} - \mu)}) + \ln(1 + e^{-\beta(\omega_{K^+} + \mu)})] . \qquad (3.111)$$

Meson fields which are modified due to the K^- condensate, can be determined by solving the following equations in the RMF approximation [54, 64],

$$m_\sigma^2 \sigma = \sum_B g_{\sigma B} n_B^S + g_{\sigma K} \sum_{\bar K} n_{\bar K} ,$$

$$m_\omega^2 \omega_0 = \sum_B g_{\omega B} n_B - g_{\omega K} \sum_{\bar K} n_{\bar K} ,$$

$$m_\rho^2 \rho_{03} = \sum_B g_{\rho B} \tau_{3B} n_B + g_{\rho K} \sum_{\bar K} \tau_{3\bar K} n_{\bar K} ,$$

$$m_\phi^2 \phi_0 = \sum_B g_{\phi B} n_B - g_{\phi K} \sum_{\bar K} n_{\bar K} . \qquad (3.112)$$

The baryon density at finite temperature is,

$$n_B = 2 \int \frac{d^3k}{(2\pi)^3} \left(\frac{1}{e^{\beta(E^* - \nu_B)} + 1} - \frac{1}{e^{\beta(E^* + \nu_B)} + 1} \right), \qquad (3.113)$$

and the scalar density is given by,

$$n_B^S = 2 \int \frac{d^3k}{(2\pi)^3} \frac{m_B^*}{E^*} \left(\frac{1}{e^{\beta(E^* - \nu_B)} + 1} + \frac{1}{e^{\beta(E^* + \nu_B)} + 1} \right), \qquad (3.114)$$

where the in-medium energies of K^- mesons is given by

$$\omega_{K^\pm} = \sqrt{(k^2 + m_K^{*2})} \pm \left(g_{\omega K} \omega_0 + g_{\rho K} \rho_{03} + g_{\phi K} \phi_0 \right) . \qquad (3.115)$$

The chemical potentials of nucleons are related to that of K^- mesons by $\mu = \mu_n - \mu_p$ [62]. For s-wave ($\mathbf{k} = 0$) condensation, the momentum dependence vanishes in $\omega_{K^\pm}$. The in-medium energy of K^- condensate decreases from its vacuum value m_K as the meson fields build up with increasing density. The onset of K^- condensate occurs when ω_{K^-} equals to its chemical potential, i.e.

$$\mu = \omega_{K^-} = m_K^* - g_{\omega K} \omega_0 - g_{\rho K} \rho_{03} - g_{\phi K} \phi_0 . \qquad (3.116)$$

This is the threshold condition for K^- condensation. As the K^- condensate does not contribute to the pressure, the pressure due to thermal (anti)kaons is $P_K = TV^{-1} \ln Z_K = -\Omega_K/V$ where Ω_K is the grand canonical thermodynamic potential for (anti)kaons.

The energy density of (anti)kaons is due to the condensate plus the thermal (anti)kaons and is given by

$$\epsilon_K = m_K^* n_K^C + \left(g_{\omega K}\omega_0 + g_{\rho K}\rho_{03} + g_{\phi K}\phi_0\right) n_K^T$$

$$+ \int \frac{d^3k}{(2\pi)^3} \left(\frac{\omega_{K^-}}{e^{\beta(\omega_{K^-}-\mu)} - 1} + \frac{\omega_{K^+}}{e^{\beta(\omega_{K^+}+\mu)} - 1}\right) .$$

The entropy density is obtained from the relation $S_K = \beta\left(\epsilon_K + P_K - \mu n_K\right)$. The entropy per baryon is given by $s = S/n_B$, where n_B is the total baryon density. The total (anti)kaon number density (n_K) is $n_K = n_K^C + n_K^T$, where n_K^C and n_K^T are the K^- condensate and thermal (anti)kaon density, respectively, as given by,

$$n_K^C = 2\left(\omega_{K^-} + g_{\omega K}\omega_0 + g_{\rho K}\rho_{03} + g_{\phi K}\phi_0\right) \bar{K} K = 2m_K^* \bar{K} K ,$$

$$n_K^T = \int \frac{d^3k}{(2\pi)^3} \left(\frac{1}{e^{\beta(\omega_{K^-}-\mu)} - 1} - \frac{1}{e^{\beta(\omega_{K^+}+\mu)} - 1}\right) . \qquad (3.117)$$

As meson-(anti)kaon couplings are not density dependent, we use those coupling constants of Sect. 3.5.2.

3.5.6 Nuclear Physics Constraints on EoS

In Sect. 3.2, we have discussed the astrophysical constraints on neutron stars. Here we discuss the nuclear physics inputs that constrain the EoS of neutron star matter. The energy per nucleon of asymmetric homogeneous nuclear matter at a density n is Taylor-expanded around the isospin asymmetry $\delta = (n_n - n_p)/n = 0$ as

$$e(n, \partial) = e(n, 0) + e_{sym}(n)\delta^2 + O(\delta^4) , \qquad (3.118)$$

where the energy per nucleon of the symmetric matter $(e(n, 0))$ and symmetry energy (e_{sym}) are expanded around the saturation density (n_0) in terms of $x = (n - n_0)/3n_0$

$$e(n, 0) = e_0 + \frac{1}{2}K_0 x^2 + \frac{1}{6}Q_0 x^3 + \cdots$$

$$e_{sym}(n) = J + Lx + \frac{1}{2}K_{sym}x^2 + \cdots , \qquad (3.119)$$

where $e_0 = e(n_0, 0)$, the incompressibility $K_0 = 9n_0^2 \frac{\partial^2 e}{\partial n^2}|_{n_0}$, the skewness parameter $Q_0 = 27n_0^3 \frac{\partial^3 e}{\partial n^3}|_{n_0}$, the symmetry energy coefficient J, the symmetry energy slope $L = 3n_0 \frac{\partial J}{\partial n}|_{n_0}$ and the symmetry incompressibility $K_{sym} = 9n_0^2 \frac{\partial^2 J}{\partial n^2}|_{n_0}$. Nuclear matter parameters correlate among each other and to the properties of nuclei and neutron star observables [204]. These nuclear matter parameters at the saturation density are determined from terrestrial experiments. The data of nuclear masses yield the nuclear matter parameters such as the saturation density and binding energy per nucleon accurately [129, 205].The incompressibility K_0 was obtained by analyzing the data of the isoscalar giant monopole resonance (ISGMR) also known as the breathing mode of nuclei and it led to the value of $K_0 = 240 \pm 20 \, \text{MeV}$ [206]. Stone et al. performed a reanalysis of ISGMR data and obtained $K_0 = 250–315 \, \text{MeV}$. [202]. On the other hand, the symmetry energy is estimated from the nuclear isovector giant dipole resonance (IVGDR). A strong correlation is found to exist between the value of the centroid IVGDR energy in spherical nuclei and the symmetry energy [207]. Furthermore, the neutron skin thickness of ^{208}Pb was found to correlate with the density dependence of the symmetry energy [208–210]. It is worth mentioning that a correlation between J and L was explored using various functional forms of E_{sym} that led to the values of $J = 29.0–32.7 \, \text{MeV}$ and $L = 40.5–61.9 \, \text{MeV}$ [198]

As introduced in Sect. 3.5.1.1, the saturation properties corresponding to different parametrizations of the RMF models with and without density dependent couplings are listed in Table 3.1 along with the maximum gravitational masses and the corresponding baryonic masses. Empirical values of nuclear matter properties are reported in the last row of Table 3.1. The EoS models shown in the table are DD2, BHB$\Lambda\phi$, SFHo, SFHx, TM1, TMA, G230a, G230b, G240a, G240b, G300a, G300b, and quark-hadron hybrid. The range of values of incompressibility of nuclear matter at the saturation density is taken from Refs. [70, 202]. It was found that the nucleon effective mass in the range $0.55 \leq m^*/m \leq 0.75$ determined the values of the symmetry energy and its slope that led to the physical solution for pure neutron matter which was compatible with the chiral effective field theory [203]. Tews et al. also obtained new bounds on the symmetry energy and its slope [199]. It is evident from the table that the symmetry energy and its slope of SFHx, TMA, and Glendenning EoS models are not in consonance with experimental values, new bounds of Tews et al. and the state of the art calculations in the chiral effective field theory [70, 199, 200]. The low value of L in the SFHx EoS model was realized in an attempt to minimize the radii of low mass neutron stars [151].

Table 3.1 The saturation properties of nuclear matter such as saturation density (n_0), the ratio (m^*/m) of the effective nucleon mass (m^*) and the bare nucleon mass (m), binding energy (BE), incompressibility (K), symmetry energy (S), and density slope of symmetry energy (L) are obtained using different parameter sets. Maximum masses of non-rotating neutron stars and the corresponding baryon mass are also mentioned here. Experimental values of nuclear matter properties at the saturation density quoted in the last row are taken from Ref. [70, 202, 203]. This is taken from Ref. [196] and reproduced by permission of the AAS. https://doi.org/10.3847/1538-4357/ab

EoS	n_0 (fm^{-3})	m^*/m	BE (MeV)	K_0 (MeV)	S (MeV)	L (MeV)	M_{max} (M$_\odot$)	M_B (M$_\odot$)
DD2	0.1491	0.56	16.02	243.0	31.67	55.04	2.42	2.89
BHB$\Lambda\phi$	0.1491	0.56	16.02	243.0	31.67	55.04	2.1	2.43
SFHo	0.1583	0.76	16.19	245.4	31.57	47.10	2.06	2.43
SFHx	0.1602	0.72	16.16	238.8	28.67	23.18	2.13	2.53
TM1	0.1455	0.63	16.31	281.6	36.95	110.99	2.21	2.30
TMA	0.1472	0.64	16.03	318.2	30.66	90.14	2.02	2.30
G230a	0.153	0.78	16.30	230.0	32.50	89.76	2.01	2.31
G230b	0.153	0.70	16.30	230.0	32.50	94.46	2.33	2.75
G240a	0.153	0.78	16.30	240.0	32.50	89.70	2.02	2.75
G240b	0.153	0.70	16.30	240.0	32.50	94.39	2.34	2.75
G300a	0.153	0.78	16.30	300.0	32.50	89.33	2.08	2.40
G300b	0.153	0.70	16.30	300.0	32.50	93.94	2.36	2.78
Hybrid	0.1491	0.56	16.02	243.0	31.67	55.04	2.05	2.39
Exp.	0.15–0.16	0.55–0.75	16.00	220–315	29.00–31.70	45.00–61.90	–	–

3.6 Tolman–Oppenheimer–Volkoff Equation and Structures of Neutron Stars

The structure of a spherically symmetric and static star is obtained by solving Einstein field equations as given by [148]

$$G^{\mu\nu} = -8\pi T^{\mu\nu} , \tag{3.120}$$

where the Einstein curvature tensor is expressed in terms of the Riemann curvature tensor as

$$G^{\mu\nu} = R^{\mu\nu} - \frac{1}{2} g^{\mu\nu} R , \tag{3.121}$$

and the perfect fluid energy-momentum tensor

$$T^{\mu\nu} = -p g^{\mu\nu} + (p + \varepsilon) u^{\mu} u^{\nu} , \tag{3.122}$$

with the four-velocity $u^{\mu} = (1, 0, 0, 0)$. Here we take geometrized unit $G = c = 1$. Next the solutions to Eq. (3.120) in the exterior and interior regions of static stars are to be found. The Schwarzschild metric is the solution of Einstein field equations outside a spherical static star. The most general solution is of the form,

$$ds^2 = e^{2\nu(r)} dt^2 - e^{2\lambda(r)} dr^2 - r^2 d\theta^2 - r^2 \sin^2 \theta d\phi^2 . \tag{3.123}$$

Here ν and λ are functions of r and $M(r)$ is the included mass inside a radius r. Finally the Einstein equations for the interior of a spherical, static and relativistic neutron star reduce to [148]

$$\frac{dp}{dr} = -\frac{[p(r) + \varepsilon(r)][M(r) + 4\pi r^3 p(r)]}{r [r - 2M(r)]} . \tag{3.124}$$

Equation (3.124) that describes the hydrostatic equilibrium is known as the TOV equation [12, 13]. The TOV equation is solved along with the following equations,

$$\frac{dM(r)}{dr} = 4\pi \varepsilon(r) r^2 ,$$

$$\frac{d\nu}{dr} = -\frac{1}{\varepsilon(r) + p} \frac{dp}{dr}, \tag{3.125}$$

to determine the radius (R) and mass $(M = M(R))$ of a neutron star. The EoS in terms of the pressure (P) as a function of energy density (ε) closes the system of equations.

We discuss the zero temperature EoS models as mentioned in Table 3.1 and the corresponding mass-radius relationships determined using the TOV equation [196]. For each EoS model, we match the low density EoS models of crusts with the EoS model of uniform matter in the core of neutron stars. For example, we add the BPS model of the outer crust along with the Negele and Vautherin model of the inner crust model for all Glendenning EoS models. The extended NSE crust model at sub-saturation densities is merged with the EoS described in the RMF models with density dependent couplings such as the DD2, BHBΛϕ including Λ hyperons and hybrid EoS undergoing a first order quark-hadron phase transition and without density dependent couplings in the case of TMA, TM1, SFHo, and SFHx involving only nucleons. It is to be noted that the same RMF parametrization that is used for the uniform matter, is adopted for the description of the interactions among unbound nucleons in the inhomogeneous matter within the extended NSE model. Figure 3.5 shows EoSs constructed within the framework of different models as described in Sect. 3.5.1 and listed in Table 3.1. It is evident from the plot that the DD2 EoS is the stiffest one among all EoS models. The beginning and end of the mixed phase of the hybrid EoS are manifested by two kinks. The hybrid EoS becomes softer with the termination of the mixed phase. The behavior of the low density part of the

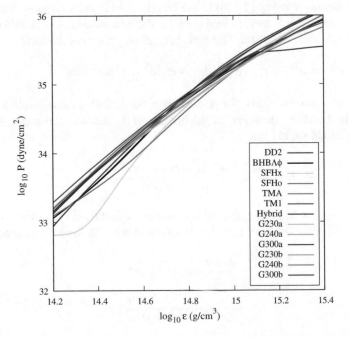

Fig. 3.5 Pressure as a function of energy density is plotted for various EoS models of Table 3.1 at zero temperature. This is taken from Ref. [196] and reproduced by permission of the AAS. https://doi.org/10.3847/1538-4357/ab

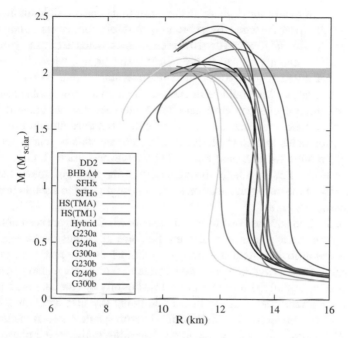

Fig. 3.6 Mass-Radius relationships corresponding to EoSs shown in Fig. 3.5 are plotted. The grey horizontal band denotes the bound on the neutron star maximum mass as obtained from the accurately measured mass 2.01±0.04 $M_{\odot}$ of PSR J0348+0432. This is taken from Ref. [196] and reproduced by permission of the AAS. https://doi.org/10.3847/1538-4357/ab6a9e

SFHx EoS is different in comparison to the other EoSs and this is attributed to the low value of the density slope of the symmetry energy of the SFHx EoS.

The mass-radius relationships of the corresponding EoSs of Fig. 3.5 are shown in Fig. 3.6. The grey band denotes the accurately measured mass 2.01±0.04 of PSR J0348+0432. It is noted that the lower effective mass in Glendenning EoS models makes the EoS very stiff leading to the higher maximum mass of neutron stars than that of higher values for incompressibility [148, 196]. Although all EoS models shown here are compatible with 2 $M_{\odot}$ neutron star, the radii vary widely. The radius corresponding to the TM1 EoS model is significantly different from others and maybe ruled out based on the recent results of the NICER.

3.7   Stable Branch of Compact Stars Beyond Neutron Star Branch

The possibility of a stable sequence of compact stars beyond the neutron star branch has generated tremendous interest in recent times. White dwarfs and neutron stars can be called the first and second families of compact stars, respectively. It is the

Fermi pressure of relativistic degenerate electrons that is responsible for the stability of white dwarfs. After the white dwarf branch, no stable compact stars can be found in the mass density interval $\sim 10^9$–10^{14} g/cm^3 as is noted from the mass-radius relationship. The stability returns in the neutron star branch when the pressure of degenerate and interacting baryons makes neutron stars stable. What is next after the neutron star branch? It would be interesting if an unstable region is followed by a stable configuration of superdense stars beyond the neutron star branch! This new stable branch of more compact stars might be called the third family. Initial studies in that direction noted that no stable branch of compact stars beyond a neutron star branch was possible for a smooth EoS [211]. However, Ulrich H. Gerlach argued that a jump in the EoS or a large discontinuity in the speed of sound beyond the central density of the maximum neutron star mass might lead to a stable third family of superdense stars [212].

Kinks in the EoS and the discontinuity in the speed of sound are common features associated with the appearance of any exotic form of matter in the high density regime. A strong first order phase transition from hadronic matter to any exotic form of matter such as hyperon, antikaon condensed or quark matter is considered to be the reason behind the formation of a third family of compact stars [64, 213–217]. It was demonstrated that the EoS undergoing the first order Bose–Einstein condensation of antikaons or a first order hadron-quark phase transition might terminate the neutron star branch due to the softening in the mixed phase and lead to the third family of compact stars made of antikaon condensates or quark matter in the core [64, 214–216]. Similar stable solutions beyond the neutron star branch were obtained in a first order phase transition from hadronic matter to hyperonic matter for certain parameters sets of hyperon-hyperon interactions [213]. The third family was also found in the calculation with the EoS involving a first order hadron-CFL quark matter phase transition [216]. It was observed that the partial overlapping mass regions of the neutron star branch and the third family branch result in non-identical stars of the same mass but distinctly different radii and compositions. Such pairs are called "neutron star twins" [214, 215]. The study of twin stars is an important area of research nowadays [218–220].

The mixed phase of a first order phase transition from the hadronic matter to an exotic form of matter is governed by the Gibbs phase rules. According to these rules, the equality of pressures and baryon chemical potentials are to be satisfied in two phases denoted by I and II as shown below,

$$P^I = P^{II}, \tag{3.126}$$

$$\mu_B^I = \mu_B^{II}, \tag{3.127}$$

where μ_B^I and μ_B^{II} and P^I and P^{II} are chemical potentials of baryon B and pressures, respectively, in two phases. Further, it is noted from the generalized β-equilibrium relation of Eq. (3.53) that two independent chemical potentials μ_n and μ_e are connected to the conservation of baryon number (n_B) and electric charge (Q) in neutron star matter. The mixed phase is described by the Gibbs phase rules for thermodynamic equilibrium along with the global baryon number and electric charge conservation laws [54, 222]. The conditions of global baryon number conservation and charge neutrality are,

$$n_B = (1 - \chi)n_B^I + \chi n_B^{II},$$
$$(1 - \chi)Q^I + \chi Q^{II} = 0, \tag{3.128}$$

where χ is the volume fraction of phase II. The total energy density in the mixed phase is given by,

$$\varepsilon = (1 - \chi)\varepsilon^I + \chi\varepsilon^{II}, \tag{3.129}$$

where ε^I and ε^{II} are energy densities in phase I and II, respectively.

Although conserved charges are shared in the mixed phase of the hadronic and unpaired quark matter, this is not the same in the case of the first order phase transition involving the CFL quark matter. It was already noted that no electrons were needed to make the CFL quark matter charge neutral. As the color charge in the bulk hadronic phase and electric charge in the CFL phase are not present, the global charge conservation is relaxed in the first order phase transition from the hadronic to the CFL matter. In this case, the local electrical charge neutrality in the hadronic phase and local color charge neutrality in the CFL phase are imposed. The mixed phase is constructed using the Gibbs phase rules and global baryon number conservation [189, 190, 216].

Figure 3.7 shows the EoS with and without the first order phase transition from the hadronic matter to antikaon condensed matter that includes K^- and $\bar{K}^0$ condensates as described in Fig. 3.4. The EoS with the antikaon condensates exhibits two kinks that correspond to the beginning and the end of the mixed phase. Those kinks lead to discontinuities in the speed of sound. This results in the termination of the neutron star branch near the upper boundary of the mixed phase due to the softening of the EoS and it is followed by a stable third family branch that contains a pure phase of K^- and $\bar{K}^0$ condensates in the core [64]. Compact stars in the third family have smaller radii ($\sim$10 km) than their counterparts in the neutron star branch. Similar results were obtained for first order hadronic matter to hyperon/unpaired quark matter phase transitions [213–215]. The extension of this study to the first order phase transition from the hadronic to the CFL quark matter also produced the stable third family branch [216]. The third families of stable compact stars and twin stars due to overlapping mass regions in the neutron star and third family branches are shown for several EoSs including hadron-quark phase transitions in Fig. 3.8 [223]. Compact stars in the third family might reach the 2 $M_\odot$ limit as evident from Fig. 3.8.

The neutron star branch and the third family branch have positive slopes, i.e. $dM/d\varepsilon_c > 0$, where ε_c is the central energy density. The positive slope of a stellar sequence is a necessary condition for the hydrostatic stability. However, the dynamical stability is tested solving eigenfrequencies of normal modes of radial

Fig. 3.7 Pressure is plotted
as a function energy density
for cases with and without
antikaon condensate using the
GM1 parameter and antikaon
potential depth of -160 MeV.
This is taken from Ref. [221]
and reproduced by permission
of the IoP Publishing

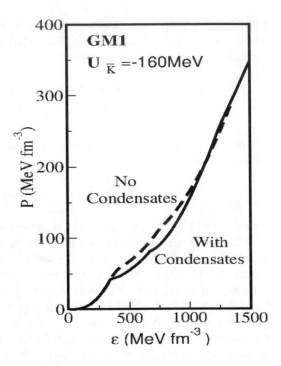

vibrations [148, 214, 215, 224–226]. The small perturbation applied on a spherically
symmetric star is given by

$$\delta r(r, t) = e^{\nu} U_n(r) e^{i\omega_n t}/r^2 , \qquad (3.130)$$

where $u_n(r)$ is the amplitude of normal modes of vibrations and ω_n represents
eigenfrequencies [148]. The eigenequation for $u_n(r)$ is a second order differential
equation that has the Strum-Liouville form [148]. One obtains the frequency
spectrum of ω_n^2 of normal radial modes where $n = 0, 1, 2 \ldots$. Any negative value
of the squared eigenfrequencies implies exponential growth of the perturbation and
the star will be unstable. In the third family branches, each configuration was found
to be stable because the squared frequency of the normal mode associated with it
was positive [64].

3.8 Rotating Neutron Stars, Moment of Inertia
and Quadrupole Moment

The majority of neutron stars discovered so far are found to rotate slowly. The fastest
neutron star has a frequency of 716 Hz [227]. Here we discuss different formalisms
employed to study slowly as well as rapidly rotating neutron stars. The effects of

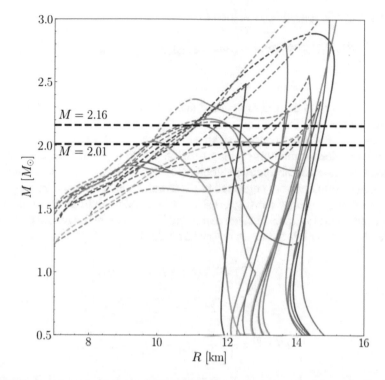

Fig. 3.8 Mass-radius relations along with twin star solutions are shown for several EoSs as described in Ref. [223]. Solid lines represent stable branches of compact stars and dashed lines imply unstable regions. Reprinted figure with permission from Most et al. [223] ©2018 by the American Physical Society (APS). https://doi.org/10.1103/PhysRevLett.120.261103

angular velocity on the structure of rigidly and slowly rotating compact stars were investigated for the first time by J. B. Hartle and K. S. Thorne [228, 229]. After the work of Hartle and Thorne, others constructed more sophisticated models of rotating neutron stars. We discuss those models and calculate the moment of inertia, quadrupole moment of rotating neutron stars.

3.8.1 Slowly Rotating Neutron Stars

A rotating star is a deformed body and the spherical symmetry is not respected in this case. The axial symmetry is retained in rotating stars. Unlike the metric in Eq. (3.123), the metric of a rotating neutron star includes some off diagonal elements. So the metric of a rigidly rotating neutron star assuming a stationary, axially symmetric spacetime is written as

$$ds^2 = e^{2\nu(r,\theta)}dt^2 - e^{2\lambda(r,\theta)}dr^2 - e^{2\mu(r,\theta)}\left[r^2 d\theta^2 + r^2 \sin^2 \theta \left(d\phi - \omega(r,\theta)dt\right)^2\right],$$

$$(3.131)$$

where ν, λ, μ, and ω depend on r and θ and $\omega(r,\theta)$ is the angular velocity of the local inertial frames.

In case of a slowly rotating neutron star, the rotation is considered as a small perturbation on the non-rotating neutron star [228]. The field equations are expressed in powers of the angular velocity (Ω). Terms higher than the second order in the angular velocity were neglected in those equations. For slowly rotating stars, the changes in the energy density, pressure, gravitational field are negligible. The metric coefficients in Eq. (3.131) are expanded in terms of the Legendre polynomial of order 2, i.e. $P_2(\cos\theta)$ as shown below [228, 229]

$$e^{2\nu(r,\theta)} = e^{2\nu}\left[1 + 2\left(h_0 + h_2 P_2(\cos\theta)\right)\right],$$

$$e^{2\lambda(r,\theta)} = e^{2\lambda}\left[1 + 2\frac{(m_0 + m_2 P_2(\cos\theta))}{r - 2M(r)}\right],$$

$$e^{2\mu(r,\theta)} = \left[1 + 2(v_2 - h_2)P_2(\cos\theta)\right],$$

$$(3.132)$$

where m_0, m_2, h_0, h_2, and v_2 are functions of r and proportional to Ω^2. These functions are obtained from Einstein field equations [228, 229]. Further ω is proportional to Ω.

The quantity denoted by

$$\bar{\omega}(r,\theta) = \Omega - \omega(r,\theta) \tag{3.133}$$

is the angular velocity of the fluid element with respect to the local inertial frame. This difference, $\omega(\bar{r},\theta)$, was shown to be function of only r and satisfy the following differential equation [228, 229]

$$\frac{1}{r^4}\frac{d}{dr}\left[r^4 j\frac{d\bar{\omega}}{dr}\right] + \frac{4}{r}\frac{dj}{dr}\bar{\omega} = 0. \tag{3.134}$$

Here $j(r)$ is defined in terms of the metric coefficients for a star as,

$$j(r) = e^{-(\nu+\lambda)} = e^{-\nu}\sqrt{(1 - 2M(r)/r)} \qquad r < R. \tag{3.135}$$

At the surface and outside the star [148, 228],

$$j(r) = 1 \qquad\qquad r \geq R, \tag{3.136}$$

and

$$\omega(r) = \Omega - \frac{2J}{r^3} \qquad\qquad r \geq R\,, \qquad (3.137)$$

where J is the angular momentum. The angular velocity of the local frames vanishes far off from the star.

Integrating Eq. (3.134) from the center of the star to the surface, one obtains

$$\left[r^4 j \frac{d\bar{\omega}}{dr}\right]_R = -\int_0^R 4r^3 \frac{dj}{dr}\bar{\omega}\,dr.$$

Further simplification using Eq. (3.137) gives the expression of the angular momentum,

$$J = -\frac{2}{3}\int_0^R r^3 \frac{dj}{dr}\bar{\omega}\,dr\,. \qquad (3.138)$$

Finally, the moment of inertia (I) is obtained from the relation $J = I\Omega$ and is given by

$$I = \frac{8\pi}{3}\int_0^R dr\,r^4 e^{-\nu}\frac{[\varepsilon(r) + P(r)]}{\sqrt{(1 - 2M(r)/r)}}\frac{[\Omega - \omega(r)]}{\Omega}\,. \qquad (3.139)$$

The quadrupole moment (Q) gives the deformation of a rotating star. Hartle and Thorne estimated Q of a slowly rotating compact star as [228, 229],

$$Q = \frac{8}{5}M^3 \times \frac{h_2 + v_2 - (J^2/MR^3)}{\frac{2M}{\sqrt{R(R-2M)}}Q_2^1(R/M - 1) + Q_2^2(R/M - 1)} + \frac{J^2}{M}\,, \qquad (3.140)$$

where $Q_m^n(x)$ is the associated Legendre function of the second kind given by

$$Q_2^1(x) = \sqrt{x^2 - 1} \times \left[\frac{3x^2 - 2}{x^2 - 1} - \frac{3}{2}x \log\left(\frac{x + 1}{x - 1}\right)\right] \qquad (3.141)$$

$$Q_2^2(x) = \frac{3}{2}(x^2 - 1)\log\left(\frac{x + 1}{x - 1}\right) - \frac{3x^3 - 5x}{x^2 - 1}\,. \qquad (3.142)$$

The functions v_2 and h_2 are obtained by solving the $\ell = 2$ equations [228, 229].

3.8.2 Fully Relativistic, Nonlinear Models of Rapidly Rotating Neutron Stars

Besides the perturbative approach of Hartle and Thorne, there are other methods where the properties of rotating neutron stars are studied by direct integration of Einstein equations. Here we discuss two such models of rotating neutron stars that do not impose any restriction on the angular velocity.

We are dealing with the rotating equilibrium models which are stationary and axisymmetric. In the formalism of Komatsu-Eriguchi-Hachisu (KEH), the metric inside the rotating neutron star has the form [230–232],

$$ds^2 = -e^{\gamma+\sigma}\, dt^2 + e^{2\alpha}\, (dr^2 + r^2 d\theta^2) + e^{\gamma-\sigma}\, r^2 \sin^2\theta\, (d\phi - \omega dt)^2\,, \quad (3.143)$$

where γ, σ, α, ω are the functions of r and θ. The stress-energy tensor of a perfect fluid is given by

$$T^{\mu\nu} = (\rho_0 + \rho_i + P)u^\mu u^\nu + P g^{\mu\nu}\,. \quad (3.144)$$

The rest energy density and internal energy density are ρ_0 and ρ_i, respectively; P is the pressure and u^μ is the four-velocity of the matter. It is assumed that there is no meridional circulation of the matter so that the four-velocity u^μ is simply a linear combination of time and angular Killing vectors.

The general relativistic field equations that determine σ, γ and ω in the KEH formalism, are written as [230–232]

$$\Delta\left[\sigma e^{\frac{\gamma}{2}}\right] = S_\sigma(r,\mu)\,, \quad (3.145)$$

$$\left(\Delta + \frac{1}{r}\partial_r - \frac{\mu}{r^2}\partial_\mu\right)\left[\gamma e^{\frac{\gamma}{2}}\right] = S_\gamma(r,\mu)\,, \quad (3.146)$$

$$\left(\Delta + \frac{2}{r}\partial_r - \frac{2\mu}{r^2}\partial_\mu\right)\left[\omega e^{\frac{\gamma}{2}-\sigma}\right] = S_\omega(r,\mu)\,, \quad (3.147)$$

where the flat-space spherical coordinate Laplacian,

$$\Delta \equiv \partial_r^2 + \frac{2}{r}\partial_r + \frac{1-\mu^2}{r^2}\partial_\mu^2 - \frac{2\mu}{r^2}\partial_\mu + \frac{1}{r^2(1-\mu^2)}\partial_\phi^2\,, \quad (3.148)$$

$\mu = \cos\theta$ and $S_\sigma, S_\gamma, S_\omega$ are the effective source terms including the nonlinear and coupling terms. The effective source terms S_σ, S_γ and S_ω have the same form as given by Eqs. (6)–(8) of Ref. [231]. The fourth field equation determining α is given by Eq. (11) of Ref. [231].

The coordinate components of the four-velocity of the matter are given by

$$u^\mu = \frac{e^{-(\sigma+\gamma)/2}}{\sqrt{1-v^2}}[1,0,0,\Omega] , \qquad (3.149)$$

where Ω is the angular velocity of the matter at infinity and v is the proper velocity of the matter with respect to a zero angular momentum observer and is defined as,

$$v = (\Omega - \omega)re^{-\upsilon}\sin\theta . \qquad (3.150)$$

The equation of hydrostatic equilibrium follows from the vanishing of the divergence of the energy-momentum tensor [231, 232].

Three elliptic field equations (3.145)–(3.147) are solved by an integral Green's function technique [230–232]. The fourth field equation simplifies to a linear ordinary differential equation [231, 232].

The Einstein field equations and the equation of hydrostatic equilibrium are solved on a discrete grid employing a combination of integral and finite difference techniques. These are implemented in the rotating neutron star (RNS) code developed by N. Stergioulas and J. L. Friedman [233, 234]. The computational domain of the problem is set by $0 \le r \le \infty$ and $0 < \mu \le 1$. A new radial coordinate s is defined by

$$r = r_e \frac{s}{1-s} , \qquad (3.151)$$

where r_e is the coordinate equatorial radius. Thus, $s = 0$ is the center; the equator is located at $s = \frac{1}{2}$ and $s = 1$ corresponds to the infinity. The RNS code is publicly available for use.[2]

Rapidly rotating star models were also numerically studied in $3+1$ dimensional space plus time formulation in the general theory of relativity [235, 236]. In this case, the four-dimensional spacetime manifold is decomposed into a family of non-intersecting space-like hyper-surfaces Σ_t having the coordinate time t as the parameter. Introducing three spatial coordinates (x^i) on each hyper-surface and the 3-metric γ_{ij} on each Σ_t, the line element in terms of the lapse function N and shift vector (β^i) is,

$$ds^2 = -N^2dt^2 + \gamma_{ij}(dx^i + \beta^i dt)(dx^j + \beta^j dt) . \qquad (3.152)$$

The spacetime symmetries and foliation in $3+1$ framework are behind the choice of coordinates in this case. The spacetime is considered to be stationary, axisymmetric and asymptotically flat. Consequently there are two commuting Killing vector fields ($\mathbf{e_0} = \frac{\partial}{\partial t}$ and $\mathbf{e_3} = \frac{\partial}{\partial \phi}$) in the chosen coordinates ($x^0 = t$,

---

[2] http://www.gravity.phys.uwm.edu/rns/.

x^1, x^2, $x^3 = \phi$). The rest coordinates ($x^1 = r, x^2 = \theta$) are chosen as spherical. Further $\beta^r = \beta^\theta = 0$, $\gamma_{r\phi} = \gamma_{\theta\phi} = 0$. Using a quasi-isotropic gauge which makes $\gamma_{r\theta} = 0$, the line element takes the form [237],

$$ds^2 = -N^2 dt^2 + A^2(dr^2 + r^2 d\theta^2) + B^2 r^2 \sin^2\theta \, (d\phi - \beta^\phi dt)^2 \,, \qquad (3.153)$$

where N, β^ϕ, A, B depend on coordinates r and θ. Finally, gravitational field equations were obtained as a set of four coupled elliptic partial differential equations including energy-momentum tensor in source terms [235].

The energy-momentum tensor of a perfect fluid describing the matter,

$$T^{\mu\nu} = (\varepsilon + P)u^\mu u^\nu + P g^{\mu\nu} \,, \qquad (3.154)$$

and the fluid log-enthalpy is

$$H = \ln\left(\frac{\varepsilon + P}{n_B m_B}\right) \,, \qquad (3.155)$$

where n_B and m_B are the baryon density and baryon rest mass, respectively.

The equation of the fluid equilibrium follows from the conservation of energy-momentum tensor

$$H(r,\theta) + \ln N - \ln\Gamma(r,\theta) = \frac{T e^{-H}}{m_B}\partial_i s - u_\phi u^t \partial_i \Omega, \qquad (3.156)$$

where Γ is the Lorentz factor of the fluid with respect to the Eulerian observer, s is the entropy per baryon in Boltzmann unit. As we consider only rigid rotation, i.e. $\Omega = $ constant, the last term disappears. It was shown that the equilibrium equation (3.156) finally became the zero temperature expression [237]

$$H(r,\theta) + \ln N - \ln\Gamma(r,\theta) = const \,. \qquad (3.157)$$

The first-integral (Eq. (3.157)) is integrable within the $3 + 1$ formalism and that reduces to Poisson-like partial differential equations. These equations are then solved numerically using the spectral scheme within the numerical library LORENE[3] [239]. The input parameters for the model are an EoS, the rotation frequency Ω and the central log-enthalpy H_c. Global properties of rotating neutron stars such as gravitational mass, angular momentum, quadrupole moment, and circumferential equatorial radius are calculated within this formalism using the asymptotic behavior of the lapse function and the component of the shift vector, A, B, and the source [235, 236]. The gravitational mass, angular momentum, and

---

[3] https://lorene.obspm.fr/.

quadrupole moment are given, respectively, as [236, 240, 241],

$$M = \frac{1}{4\pi} \int \sigma_{\ln N} \, r^2 \, \sin^2 \theta \, dr \, d\theta \, d\phi \tag{3.158}$$

$$J = \int A^2 \, B^2 \, (E + p) \, U \, r^3 \, \sin^2 \theta \, dr \, d\theta \, d\phi \, . \tag{3.159}$$

$$Q = -M_2 - \frac{4}{3} \left(b + \frac{1}{4} \right) M^3 \, , \tag{3.160}$$

where

$$M_2 = -\frac{3}{8\pi} \int \sigma_{\ln N} \left(\cos^2 \theta - \frac{1}{3} \right) r^4 \, \sin^2 \theta \, dr \, d\theta \, d\phi \tag{3.161}$$

Here, $\sigma_{\ln N}$ is given by the right hand side of Eq. (3.19) of [235], U is the fluid four-velocity, $E = \Gamma^2(\varepsilon + p) - p$ and $\Gamma = (1 - U^2)^{-1/2}$ and b is defined by Eq. (3.37) of [241]. The moment of inertia of the rotating star is defined as,

$$I := \left| \frac{J}{\Omega} \right| \, . \tag{3.162}$$

We now discuss results on the moment of inertia and quadrupole moment. These findings are based on slowly rotating neutron stars with spin frequency 100 Hz using the numerical library LORENE. This investigation was carried out using some representative EoS models of Table 1.1 [238]. The BHB$\Lambda K^-\phi$ EoS includes a K^- condensate. The dependence of the moment of inertia on the neutron star mass is shown in Fig. 3.9. It is observed that the moments of inertia of neutron stars including strange matter are lower compared with the results of nuclear matter at higher neutron star masses. As the mass (1.337 $M_\odot$) of pulsar A in PSR J0737-3039 is accurately known, the value of the moment of inertia could be predicted from Fig. 3.9. Similarly, the quadrupole moments of rotating neutron stars having a spin frequency of 100 Hz were evaluated using the numerical library LORENE. The behavior of the calculated quadrupole moment with respect to the Kerr value is exhibited as a function of neutron star mass in Fig. 3.10. It is found that the quadrupole moment decreases as the neutron star mass decreases and approaches the Kerr value around the maximum neutron star mass. The result with the stiffest EoS of the nuclear matter is the closest to the Kerr solution. Similar results were found by others [242–244].

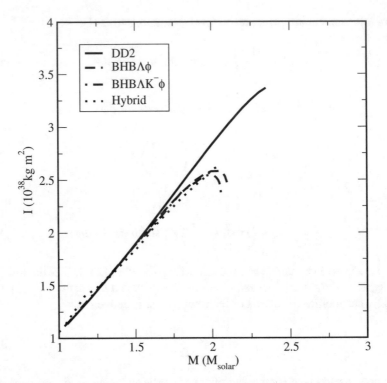

Fig. 3.9 Moment of inertia of rotating neutron star at 100 Hz is shown as a function of mass for different EoSs. This is taken from Ref. [238] and reproduced with kind permission of the European Physical Journal

3.9   Neutron Star Matter in Strongly Quantizing Magnetic Fields

Pulsars with strong surface magnetic fields $\sim 10^{12}$ G are normally found to exist in nature. However, it was observed that soft gamma-ray repeaters (SGRs) and anomalous x-ray pulsars (AXPs) had even stronger surface magnetic fields $\geq 10^{15}$ G [246, 247]. These SGRs and AXPs are believed to be the best candidates for a new class of neutron stars with very intense magnetic fields and are known as magnetars [248, 249]. It is inferred from the virial theorem that the interior magnetic field could be much higher than the surface field [250]. For a typical neutron star of mass 1.5 $M_\odot$ and radius 15 km, the interior field could be as high as $\sim 10^{18}$ G. Such strong fields would quantize the motion of charged particles in the neutron star interior. Relativistic effects are important when the cyclotron energy of the charged particle is equal to its rest mass. This leads to the critical field of electrons $B_c^e = 4.414 \times 10^{13}$ G and that for protons $B_c^p = 1.487 \times 10^{20}$ G. In this section, we discuss the influence of strongly quantizing magnetic fields on the neutron star matter in the crust and the core.

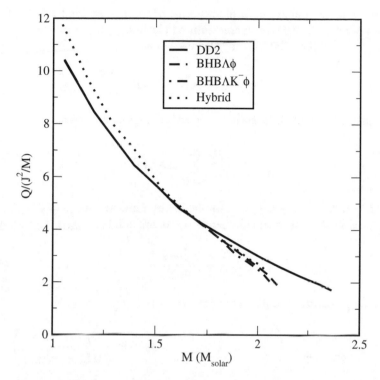

Fig. 3.10 Quadrupole moment of rotating neutron star at 100 Hz is shown as a function of mass for different EoSs. This is taken from Ref. [238] and reproduced with kind permission of the European Physical Journal

3.9.1   Magnetized Neutron Star Crusts

The compositions and EoSs in the outer and inner crusts in presence of strong magnetic fields $\sim 10^{16}$ G were widely investigated. Lai and Shapiro studied β-equilibrated and charge neutral noninteracting matter in strong magnetic fields at the sub-saturation densities [251]. This calculation was an extension of the BPS model of the outer crust as described in the Sect. 3.4.4.1 in magnetic fields [245, 251, 252]. In presence of magnetic fields, the electron motion perpendicular to the direction of the field is quantized into Landau orbitals. The energy eigenvalue of relativistic electrons in a quantizing magnetic field (B_m) is given by

$$E_e(\nu, p_z) = \left[p_z^2 + m_e^2(1 + 2\nu B^*) \right]^{1/2} , \qquad (3.163)$$

where p_z is the z-component of momentum, $B^* = B_m/B_c^e$, and ν is the Landau quantum number. The Fermi momentum of electrons, p_{f_e}, is obtained from the electron chemical potential (μ_e) as given by,

$$\mu_e = \left[p_{f_e}(\nu)^2 + m_e^2(1 + 2\nu B^*) \right]^{1/2} . \tag{3.164}$$

The number density of electrons in a magnetic field is calculated as

$$n_e = \frac{e B_m}{2\pi^2} \sum_{\nu=0}^{\nu_{max}} g_\nu p_{f_e}(\nu) . \tag{3.165}$$

Here the spin degeneracy is $g_\nu = 1$ for the lowest Landau level ($\nu = 0$) and $g_\nu = 2$ for all other levels. The maximum Landau quantum number (ν_{max}) is given by

$$\nu_{max} = \frac{\mu_e^2 - m_e^2}{2 e B_m} . \tag{3.166}$$

The energy density and pressure of electrons are,

$$\varepsilon_e = \frac{e B_m}{4\pi^2} \sum_{\nu=0}^{\nu_{max}} g_\nu \left(p_{f_e}(\nu)\mu_e + (m_e^2 + 2 e B_m \nu) \ln \frac{p_{f_e}(\nu) + \mu_e}{\sqrt{(m_e^2 + 2 e B_m \nu)}} \right) ,$$

$$P_e = \frac{e B_m}{4\pi^2} \sum_{\nu=0}^{\nu_{max}} g_\nu \left(p_{f_e}(\nu)\mu_e - (m_e^2 + 2 e B_m \nu) \ln \frac{p_{f_e}(\nu) + \mu_e}{\sqrt{(m_e^2 + 2 e B_m \nu)}} \right) .$$

$$\tag{3.167}$$

These expressions of Eq. (3.167) are inserted in the Gibbs free energy of the BPS model.

The equilibrium nuclei and EoS of the outer crust were modified due to the Landau quantization of electrons in strong magnetic fields, $\sim 10^{16}$ G. It was noted that the neutron drip point was delayed to the higher density compared with the field free case [252].

Similarly, the inner crust model based on the BLV prescription in Sect. 3.4.4.2 was extended to the magnetic case [145]. It was found that the electron number density and energy density were most affected in strongly quantizing magnetic fields 10^{17} G and the β-equilibrium condition was modified compared with the field free case. In this case, the Fermi momentum of electrons, p_{f_e}, is obtained from the average electron chemical potential in a magnetic field

$$\mu_e = \left[p_{f_e}(\nu)^2 + m_e^2(1 + 2\nu B^*) \right]^{1/2} - \langle V^c(r) \rangle , \tag{3.168}$$

where $\langle V^c(r) \rangle$ denotes the average single particle Coulomb potential as obtained from Eq. (3.18). The significant enhancement of the electron number density in magnetic fields $\geq 10^{17}$ G due to the population of the zeroth Landau level led to the increase in the proton fraction due to the charge neutrality condition. Less number of neutrons was found to drip out of a nucleus in presence of strong fields than the zero field case. Consequently, this yielded in a larger mass number and proton number of a nucleus in presence of a magnetic field of $>10^{17}$ G, compared to the zero field case. This is shown for the calculation with the Skyrme interaction Sk272 for $B_m = 0$ and $B^* = 10^4$ in Fig. 3.11. The energy per nucleon of the system was found to decrease in magnetic fields $\geq 10^{17}$ G [145].

Magnetized crusts were used in the investigations of quasi-periodic oscillations in magnetars [253]. In this context, the shear modulus of crusts which is sensitive to the crustal compositions was enhanced in strong fields of 10^{17} G or more. It was also noted that first overtones of magneto-elastic modes confined to the crust were strongly affected in strong magnetic fields [253].

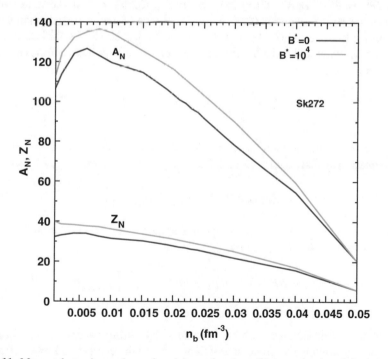

Fig. 3.11 Mass and atomic numbers of nuclei as a function of baryon density of the inner crust for magnetic fields $B_m = 0$ and 4.414×10^{17} G. This is taken from Ref. [245] and reproduced by permission of the IoP publishing

3.9.2   Dense Matter in Strong Magnetic Fields

The impact of strong magnetic fields on the composition and EoS of dense nuclear matter in neutron stars was first investigated within the RMF model in a seminal paper by Chakraborty, Bandyopadhyay and Pal [254]. Here we describe the strong magnetic field effects on the β-equilibrated dense nuclear matter. In a uniform magnetic field B_m along z-axis, the Lagrangian density (Eq. (3.33)) becomes

$$\mathcal{L}_B = \sum_{b=n,p} \bar{\psi}_b \left[i\gamma_\mu D^\mu - m_b + g_{\sigma b}\sigma - g_{\omega b}\gamma_\mu \omega^\mu - g_{\rho b}\gamma_\mu \boldsymbol{\tau} \cdot \boldsymbol{\rho}^\mu \right] \psi_b$$

$$+ \frac{1}{2}(\partial^\mu \sigma)^2 - \frac{1}{2}m_\sigma^2 \sigma^2 - U(\sigma) - \sum_{k=\omega,\rho} \left[\frac{1}{4}\left(\partial_\mu V_\nu^k - \partial_\nu V_\mu^k \right)^2 - \frac{1}{2}m_k^2 (V_\mu^k)^2 \right]$$

$$- F^{\mu\nu} F_{\mu\nu} \ . \tag{3.169}$$

Here, $D^\mu = \partial^\mu + iqA^\mu$, $U(\sigma)$ is given by Eq. (3.35), $F^{\mu\nu}$ is the field strength tensor of the electromagnetic (EM) field. The choice of gauge corresponding to the constant B_m along z-axis is $A_0 = 0$, $\mathbf{A} \equiv (0, xB_m, 0)$. The EM field is assumed to be externally generated and it has frozen field configurations. The lepton Lagrangian density is given by,

$$\mathcal{L}_L = \sum_l \bar{\psi}_l \left[i\gamma_\mu D^\mu - m_l \right] \psi_l \ . \tag{3.170}$$

We compute the equations of motion for meson fields in the mean field approximation as discussed in Sect. 3.5.1. The general solution for protons is

$$\psi_p(\mathbf{r}) \propto e^{-i\epsilon^H t + ip_y y + ip_z z} f_{p_y,p_z}(x) \ , \tag{3.171}$$

where $f_{p_y,p_z}(x)$ represents the 4-component spinor. The Dirac equation for protons in a magnetic field is then given by

$$\left[-i\alpha_x \partial/\partial x + \alpha_y (p_y - qB_m x) + \alpha_z p_z + \beta m_p^* + U_{0,p}^H \right] f_{p_y,p_z}^{(r)}(x) = \epsilon^H f_{p_y,p_z}^{(r)}(x) \ . \tag{3.172}$$

The equation of motion for neutrons is obtained by setting the charge $q = 0$ in Eqs. (3.169) and following the procedure of Sect. 3.5 and the corresponding solution is a plane wave. The interaction energy U_0^H for i-th baryon is given by

$$U_{0;i}^H = g_{\omega i}\omega_0 + g_{\rho i}\tau_{3i}\rho_{03} \ . \tag{3.173}$$

The equation of motion for free leptons (electrons and muons) in a homogeneous magnetic field is obtained by putting $U_{0,p}^H = 0$ in Eq. (3.172).

Since we deal with the cold neutron star matter, only positive energy spinors are considered. These have the following forms [255],

$$
f^{(1)}_{p_y, p_z}(x) = N_\nu \begin{pmatrix} (\epsilon^H_\nu + p_z) I_{\nu; p_y}(x) \\ -i \sqrt{2\nu q B_m} I_{\nu-1; p_y}(x) \\ -m^*_p I_{\nu; p_y}(x) \\ 0 \end{pmatrix} ,
\tag{3.174}
$$

$$
f^{(2)}_{p_y, p_z}(x) = N_\nu \begin{pmatrix} 0 \\ -m^*_p I_{\nu-1; p_y}(x) \\ -i \sqrt{2\nu q B_m} I_{\nu; p_y}(x) \\ (\epsilon^H_\nu + p_z) I_{\nu-1; p_y}(x) \end{pmatrix} ,
\tag{3.175}
$$

where $N_\nu = 1/\sqrt{2\epsilon^H_\nu (\epsilon^H_\nu + p_z)}$, the effective proton mass $m^*_p = m_p - U^H_S$, and the effective Hartree energy, $\epsilon^H_\nu = \epsilon^H - U^H_{0,p} = (p_z^2 + m^{*2}_p + 2\nu q B_m)^{1/2}$. The Landau quantum number (ν) takes all possible positive integer values including zero. The function $I_{\nu; p_y}(x)$ is similar in form as in Ref. [255],

$$
I_{\nu; p_y}(x) = \left(\frac{q B_m}{\pi}\right)^{\frac{1}{4}} \frac{1}{2^{\nu/2}\sqrt{\nu!}} e^{-\frac{\zeta^2}{2}} H_\nu(\zeta) ,
$$

$$
\zeta = \sqrt{q B_m} \left(x - \frac{p_y}{q B_m}\right) .
\tag{3.176}
$$

The self-energy of σ meson is given by

$$
U^H_S = (g_\sigma/m_\sigma)^2 n_S .
\tag{3.177}
$$

The total scalar density is $n_S = n^{(n)}_S + n^{(p)}_S$, where

$$
n^{(n)}_S = \frac{m^*_n}{2\pi^2} \left[\mu^*_n k_{F_n} - m^{*2}_n \ln \left\{ \frac{\mu^*_n + k_{F_n}}{m^*_n} \right\} \right] ,
\tag{3.178}
$$

$$
n^{(p)}_S = \frac{m^*_p q B_m}{2\pi^2} \sum_{\nu=0}^{\nu^{(p)}_{max}} g_\nu \ln \left[\frac{\mu^*_p + k_p(\nu)}{\left(m^{*2}_p + 2\nu q B_m\right)^{1/2}} \right] ,
\tag{3.179}
$$

with the Fermi momentum of neutrons $k_{F_n} = \sqrt{\mu_n^{*2} - m_n^{*2}}$, and for protons $k_p(\nu) = \sqrt{\mu_p^{*2} - m_p^{*2} - 2\nu q B_m}$. The total baryon number density is $n_B = n_n + n_p$ and

$$n_n = \frac{k_{F_n}^3}{3\pi^2}; \qquad n_p = \frac{q B_m}{2\pi^2} \sum_{\nu=0}^{\nu_{\max}^{(p)}} k_p(\nu) , \qquad (3.180)$$

where $\nu_{\max}$ is the largest integer not exceeding $(\mu_p^{*2} - m_p^{*2})/(2q B_m)$, and the effective proton chemical potential μ_p^* is ϵ_ν^H at the Fermi surface. The Landau level degeneracy factor g_ν is 1 for $\nu = 0$ and 2 for $\nu > 0$. In case of $B_m \geq B_c^e$, the charge neutrality condition, $n_p = n_e$, leads to

$$\sum_{\nu=0}^{\nu_{\max}^{(p)}} g_\nu k_p(\nu) = \sum_{\nu=0}^{\nu_{\max}^{(e)}} g_\nu k_e(\nu) , \qquad (3.181)$$

where $k_e(\nu) = \sqrt{\mu_e^2 - m_e^2 - 2\nu q B_m}$.

The total energy density of β-equilibrated and charge neutral nuclear matter is,

$$\begin{aligned}
\varepsilon = & \frac{1}{2}m_\sigma^2\sigma^2 + U(\sigma) + \frac{1}{2}m_\omega^2\omega_0^2 + \frac{1}{2}m_\rho^2\rho_{03}^2 \\
& + \frac{1}{8\pi^2}\left(2k_{F_n}\mu_B^{*3} - k_{F_n}m_n^{*2}\mu_B^* - m_n^{*4}\ln\frac{k_{F_n} + \mu_n^*}{m_n^*}\right) \\
& + \frac{q B_m}{(2\pi)^2}\sum_{\nu=0}^{\nu_{\max}} g_\nu \left\{ k_p(\nu)\mu_p^* + (m_p^{*2} + 2\nu q B_m)\ln\frac{k_p(\nu) + \mu_p^*}{\sqrt{(m_p^{*2} + 2\nu q B_m)}} \right\} \\
& + \frac{q B_m}{(2\pi)^2}\sum_{l=e,\mu}\sum_{\nu=0}^{\nu_{\max}} \left\{ k_l(\nu)\mu_l + (m_l^2 + 2\nu q B_m)\ln\frac{k_l(\nu) + \mu_l}{\sqrt{(m_l^2 + 2\nu q B_m)}} \right\} \\
& + \frac{B_m^2}{8\pi^2} ,
\end{aligned} \qquad (3.182)$$

where $k_l(\nu) = \sqrt{\mu_l^2 - m_l^2 - 2\nu q B_m}$.

Similarly, the total pressure of the nuclear matter is given by,

$$
P = -\frac{1}{2}m_\sigma^2 \sigma^2 - U(\sigma) + \frac{1}{2}m_\omega^2 \omega_0^2 + \frac{1}{2}m_\rho^2 \rho_{03}^2 + \frac{1}{3\pi^2} \int_0^{k_{Fn}} \frac{k^4 \, dk}{(k^2 + m_n^{*2})^{1/2}}
$$

$$
+ \frac{qB_m}{(2\pi)^2} \sum_{\nu=0}^{\nu_{\max}} \left\{ k_p(\nu)\mu_p^* - (m_p^{*2} + 2\nu q B_m) \ln \frac{k_p(\nu) + \mu_p^*}{\sqrt{(m_p^{*2} + 2\nu q B_m)}} \right\}
$$

$$
+ \frac{qB_m}{(2\pi)^2} \sum_{l=e,\mu} \sum_{\nu=0}^{\nu_{\max}} \left\{ k_l(\nu)\mu_l - (m_l^2 + 2\nu q B_m) \ln \frac{k_l(\nu) + \mu_l}{\sqrt{(m_l^2 + 2\nu q B)}} \right\}
$$

$$
+ \frac{B_m^2}{8\pi^2} . \tag{3.183}
$$

The parameters of Eq. (3.169) were taken from Glendenning and Moszkowski as well as Horowitz and Serot [148, 254, 256, 257]. We first focus on the calculations with the parameters of Horowitz and Serot. The impact of strong magnetic fields on the effective nucleon mass for this parameter set was significantly reduced compared with the zero field case when the magnetic field exceeded the critical field for protons, i.e. $B_m > B_c^p$ [254]. The effects of quantizing magnetic fields were further investigated on the energy per particle and proton fraction. When both electrons and protons are strongly quantized with $B_m = B_c^p$, the β-equilibrated and charge neutral nuclear matter becomes strongly bound than the field free case [254]. Consequently, the EoS in such a strong field is softer than that of $B_m = 0$. Similarly the proton fraction is enhanced in this quantizing field. When $B_m \geq B_c^e$, but appreciably smaller than B_c^p, large number of Landau levels are populated and the relation (Eq. (3.181)) corresponds to the field free case. However, when the field is B_c^p, it significantly affects electrons so that $\nu_{\max}$ is small (≈ 0), and protons are also affected. As a consequence of the charge neutrality, $\nu_{\max}^{(p)} = \nu_{\max}^{(e)} = 0$ and $k_F^{(p)} = k_F^{(e)}$ [251]. This enhancement is crucial for the onset of the direct URCA process [258, 259]. Qualitatively similarly results were obtained with the parameters of Glendenning and Moszkowski [256]. So far we discussed the energy density and pressure of the matter. The contribution due to the EM field in the energy density and the pressure is given by $B^2/8\pi^2 = 4.814 \times 10^{-8} B^{*2}$. It was noted that this contribution dominated over the matter pressure around the saturation density for $B^* > 10^4$ and in the core of the neutron star for $B^* > 10^5$ [256]. An ansatz was proposed for the variation of the magnetic field from the crust to the core as given by [260],

$$
B_m(n_b/n_0) = B_m^{\text{surf}} + B_0 \left[1 - \exp\left\{ -\beta(n_B/n_0)^\gamma \right\} \right], \tag{3.184}
$$

where the parameters are chosen to be $\beta = 0.01$ and $\gamma = 3$. The maximum field prevailing at the center is taken as $B_0 = 5 \times 10^{18} \text{G}$ and the surface field is

$B_m^{surf} \simeq 10^{14}$G. Recently, a universal parametrization of the magnetic field profile as a function of the dimensionless stellar radius was generated based on the full numerical calculation of the magnetic field distribution [261].

It was observed that the anomalous magnetic moments of nucleons might also contribute appreciably to the EoS for $B^* > 10^5$ [256]. In presence of strongly quantizing magnetic fields, the EoS becomes stiff due to the complete spin polarization of neutrons which might dwarf the softening of the Landau quantized EoS. The model of β-equilibrated and charge neutral nuclear matter in strongly quantizing magnetic fields was applied in the hadron-quark phase transition, rapid cooling of magnetized neutron stars, and the strange matter such as hyperon matter and Bose–Einstein condensate of antikaons [260, 262–267].

3.10   EoS Tables for Supernova and Binary Neutron Star Merger Simulations

We have already discussed the microphysical inputs such as the EoS of matter and neutrino interactions in the medium in the context of core collapse supernovae (CCSN) as mentioned in Chap. 2. We have also noted the importance of EoS tables in binary neutron star (BNS) merger simulations in numerical relativity, extensively studied after the discovery of gravitational waves from the first BNS merger event GW170817 [268]. Here we describe various EoS tables based on nonrelativistic as well as relativistic models.

The EoS tables of matter are constructed in the parameter space of density, temperature and positive charge fraction. R. G. Wolff and W. Hildebrandt computed the first EoS table for CCSN [269]. This was followed by the Lattimer and Swesty (LS) EoS and Shen EoS tables taking account into account all possible compositions of matter ranging from low to very high values of density, temperature, and proton fraction relevant to CCSN and BNS mergers [270, 271]. The LS EoS was constructed using the compressible liquid drop model for inhomogeneous matter and the nonrelativistic Skyrme interaction for the uniform matter whereas the Shen EoS table was computed within the framework of the RMF model. Furthermore, the LS and Shen EoS tables adopted the single nucleus approximation. Both EoS tables are being widely used in supernova and BNS merger simulations.

Many EoS tables were generated incorporating the evolving knowledge of nuclear experimental data and neutron star observations [146, 151, 272–280]. The compositions of these EoSs were mainly nucleons. There were different treatments for the inhomogeneous and uniform matter in those cases. Majority of those EoS tables were compatible with 2 $M_\odot$ neutron stars. Further development of EoS tables including hyperon and quark matter also took place [165, 281–287]. However, none of those was compatible with the heaviest neutron star observed so far. But there are EoS tables with exotic matter in agreement with neutron star observations. The quark-hadron hybrid EoS of T. Fischer et al. is consistent with 2 $M_\odot$ neutron stars [288]. The only hyperon EoS that satisfies the 2 $M_\odot$ neutron star limit and used extensively for supernova and BNS merger simulations, is the BHB$\Lambda\phi$ EoS [158]. Recently this EoS table has been extended to include the Bose–Einstein condensate of K^- mesons [289, 290]. These EoS tables with only K^- condensate and Λ hyperons including K^- condensate are denoted as DD2K^- and BHB$\Lambda K^-\phi$, respectively.

Here we describe the EoS tables based on the finite temperature density dependent model (DD2) of Sect. 3.5.4 and computed over the range of values for density $(10^{-12} \sim 1\,\text{fm}^{-3})$, temperature $(0.1–158.48\,\text{MeV})$ and positive charge fraction $(0.01–0.60)$. The NSE model of Sect. 3.4.4.3 was adopted for the calculation of inhomogeneous matter below the saturation density. The DD2 model was extended to include hyperons and antikaons in uniform matter as discussed in Sect. 3.5.1. The EoS in the nonhomogeneous and uniform matter was matched around the saturation density comparing free energies in two cases under the condition that there would not be any population of hyperons and /or (anti)kaons at low densities and temperatures. Furthermore, thermodynamic consistency checks were performed on the EoS tables [158, 289, 290]. These are unified EoSs in the sense that the same RMF model is used for the description baryon-baryon interactions both in the inhomogeneous and uniform matter. First six EoS models of Table 3.1 are examples of the unified EoSs [70]. These EoS tables are available from the CompOSE.[4]

The composition of matter is shown in Fig. 3.12 for the BHB$\Lambda\phi$ EoS table. Mass fractions of different particle species are plotted with mass density for $T = 1, 10,$ and $100\,\text{MeV}$ and positive charge fraction $Y_q = 0.1, 0.3$ and 0.5. Heavy nuclei ($Z \geq 6$) denoted by X_A are populated at lower densities and temperatures whereas light nuclei (X_a) appear at lower densities and higher temperatures. However, no nucleus is found to exist for temperature above $\sim 50\,\text{MeV}$. All nuclei dissolve to form a uniform nuclear matter around the saturation density which is $n_0 = 0.149\,\text{fm}^{-3}$ in this model. It is found that Λ hyperons appear at $2.22\,n_0$ when the chemical potential of Λ hyperons is equal to the neutron chemical potential. In the case of the BHB$\Lambda K^-\phi$ EoS table, the early appearance of Λ hyperons delays the formation of the antikaon condensate [290].

The pressure as a function of baryon mass density is plotted for the DD2, BHBΛ and BHB$\Lambda\phi$ EoS tables in Fig. 3.13. Results are demonstrated for $T = 1, 10,$ and $100\,\text{MeV}$ and $Y_p = 0.1, 0.3,$ and 0.5. The pressure only includes hadronic contributions. The pressure does not differ at all for three EoS tables at lower densities as it is evident from all the panels of Fig. 3.13. The hyperon EoSs become softer than the DD2 EoS at higher densities. It is noted that the BHB$\Lambda\phi$ EoS is stiffer compared with the BHBΛ EoS because of the repulsive hyperon-hyperon interaction mediated by ϕ mesons in the former case [158]. The appearance of the antikaon condensate along with Λ hyperons makes the BHB$\Lambda K^-\phi$ softer further [290].

---

[4] https://compose.obspm.fr.

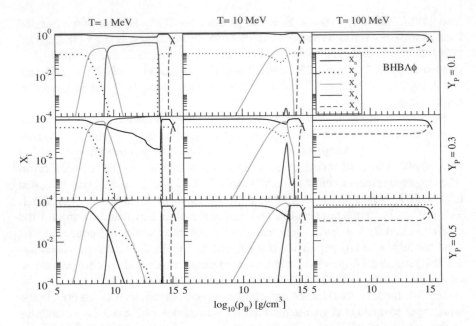

Fig. 3.12 Mass fractions of different particle species are plotted against the mass density for the BHBΛφ EoS model with $T = 1$, 10, and 100 MeV and $Y_p = 0.1$, 0.3, and 0.5. This is taken from Ref. [158] and reproduced by permission of the American Astronomical Society (AAS). https://doi.org/10.1088/0067-0049/214/2/22

The no-show of a neutron star in the supernovae explosion in 1987 (SN1987A) was attributed to the appearance of some form of exotic matter such as hyperons, antikaon condensate or quarks during the evolution of the protoneutron star (PNS) [292–295]. This led to the metastability of the PNS and its collapse to a black hole. This problem was studied in CCSN simulations using EoSs which were not compatible with 2 $M_\odot$ neutron stars [293, 295, 296]. The long duration (~3 s) evolution was investigated in the one dimensional general relativistic (GR1D) CCSN model using the BHBΛφ EoS [297, 298]. The CCSN simulations were performed with a 23 $M_\odot$ progenitor denoted by s23WH07 and 20 $M_\odot$ progenitor as s20WH07 [291]. The long duration evolutions of PNSs are shown in Fig. 3.14. No onset of the metastability in the PNSs due to the loss of the thermal support and neutrino pressure during the cooling phase over a few seconds was observed. The PNSs remained stable over 3 s. Those PNSs might evolve into cold neutron stars. In the case of s23WH07 progenitor model, the PNS would evolve to a 2 $M_\odot$ neutron star.

We discuss more applications of EoS tables in the context of BNS merger simulations in Chap. 4.

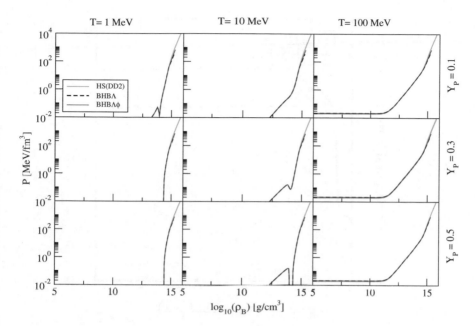

Fig. 3.13 Pressure as a function of mass density is shown for the BHBΛφ EoS model with $T = 1$, 10, and 100 MeV and $Y_p = 0.1$, 0.3, and 0.5. This is taken from Ref. [158] and reproduced by permission of the American Astronomical Society (AAS). https://doi.org/10.1088/0067-0049/214/2/22

Appendix 1

The Dirac field of Eq. (3.42) is expanded in terms of plane wave solutions with positive and negative energies as

$$\psi_B(x,t) = \int \frac{d^3 p}{(2\pi)^3} \frac{1}{\sqrt{2E_p}} \sum_s \left(a_p^s u^s(p) e^{-ip\cdot x} + b_p^{\dagger s} v^s(p) e^{ip\cdot x} \right) \quad (3.185)$$

$$\psi_B^\dagger(x,t) = \int \frac{d^3 p'}{(2\pi)^3} \frac{1}{\sqrt{2E_{p'}}} \sum_{s'} \left(a_{p'}^{\dagger s'} u^{\dagger s'}(p') e^{ip'\cdot x} + b_{p'}^{s'} v^{\dagger s'}(p') e^{-ip'\cdot x} \right)$$

$$(3.186)$$

Here $u^{1,2}(p)$ and $v^{1,2}(p)$ are positive and negative energy spinors, respectively; the annihilation operator (a_p^s) of particles is multiplied by the positive energy spinor and the creation operator $(b_p^{\dagger s})$ of antiparticles multiplies the negative energy spinor and summations are taken over spins s and s'.

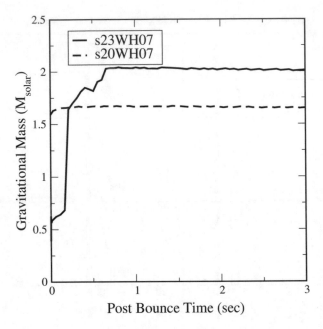

Fig. 3.14 Gravitational masses of PNSs are plotted with postbounce time for the BHBΛφ EoS model with progenitor mass models s20WH07 and s23WH07. This is taken from Ref. [291] and reproduced by permission of the Springer

The equal time anti-commutation relations for the Dirac field operators are,

$$\{\psi_B(x,t), \psi_B^\dagger(x',t)\} = \delta^3(x-x')$$

$$\{\psi_B(x,t), \psi_B(x',t)\} = \{\psi_B^\dagger(x,t), \psi_B^\dagger(x',t)\} = 0.$$

The creation and annihilation operators for particles and antiparticles follow the anti-commutation relations,

$$\{a_p^s, a_{p'}^{\dagger s'}\} = \{b_p^s, b_{p'}^{\dagger s'}\} = (2\pi)^3 \delta^3(p-p')\delta_{ss'},$$

$$\{a_p^s, a_{p'}^{s'}\} = \{a_p^{\dagger s}, a_{p'}^{\dagger s'}\} = 0,$$

$$\{b_p^s, b_{p'}^{s'}\} = \{b_p^{\dagger s}, b_{p'}^{\dagger s'}\} = 0. \tag{3.187}$$

The operator, $\bar{\psi}\psi = \psi^{\dagger}(\gamma^0)^2\psi = \psi^{\dagger}\psi$, is written as,

$$\psi_B^{\dagger}\psi_B = \int \frac{d^3p'}{(2\pi)^3} \frac{d^3p}{(2\pi)^3} \frac{1}{2\sqrt{E_p E_{p'}}} \times$$

$$\sum_s \sum_{s'} \left(a_{p'}^{\dagger s'} a_p^s u^{\dagger s'}(p')u^s(p)e^{i(p'-p)x} + a_{p'}^{\dagger s'} b_p^{\dagger s} u^{\dagger s'}(p')v^s(p)e^{-i(p+p')x} \right.$$

$$\left. + b_{p'}^{s'} a_p^s v^{\dagger s'}(p')u^s(p)e^{i(p+p')x} + b_p^s b_{p'}^{\dagger s'} v^s(p)v^{\dagger s'}(p')e^{-i(p'-p)x} \right).$$

Using the orthogonality relations,

$$u^{\dagger s'}(p)v^s(p) = v^{\dagger s'}(p)u^s(p) = 0; \qquad v^{\dagger s'}(p)v^s(p) = 2E_p\delta_{s's};$$

$$u^{\dagger s'}(p)u^s(p) = 2E_p\delta_{s's}.$$

The baryon number density operator is given by,

$$\frac{1}{V}\int d^3x \ \psi_B^{\dagger}\psi_B = \int \frac{d^3p}{(2\pi)^3} \sum_s \left[a_p^{\dagger s}a_p^s + b_p^s b_p^{\dagger s} \right], \tag{3.188}$$

where V is the volume.

After normal order of the operator of Eq. (3.188) and neglecting the constant term, it simplifies to,

$$\frac{1}{V}\int d^3x \ : \psi_B^{\dagger}\psi_B : \ = \int \frac{d^3p}{(2\pi)^3} \sum_s [a_p^{\dagger s}a_p^s - b_p^{\dagger s}b_p^s]. \tag{3.189}$$

The many-body system is considered to be in the ground state, denoted by $|F\rangle$. As the matter is degenerate, all levels up to the Fermi level are filled up. Therefore,

$$a_p^{\dagger s}a_p^s|F\rangle = N_p|F\rangle$$

N_p is the occupation number of the state. The operator $a_p^{\dagger s}a_p^s$ acting on $|F\rangle$ gives the number of fermions with momentum "p" and spin "s." Since, the matter is cold and confined in the ground state, the antiparticles are neglected. The value of N_p is either 1 or 0 depending on whether the state with momentum p is filled up or empty. The baryon number density can be written as

$$n_B = \frac{1}{V}\int d^3x \ \langle F| : \psi_B^{\dagger}\psi_B : |F\rangle = \int_0^{p_{F_B}} \frac{2}{(2\pi)^3}d^3p. \tag{3.190}$$

The factor "2" arises due to the spin degeneracy.

Using the above result in the ω meson field equation (Eq. (3.41)), we obtain

$$\omega_0 = \sum_B \frac{g_{\omega B}}{m_\omega^2} n_B .$$ (3.191)

Similarly, the calculation of the quantity, $\bar{\psi}\psi = \psi^\dagger \gamma^0 \psi$ in the σ meson field in Eq. (3.41), leads to the computation of $\sum_s u^{\dagger s}(p)\gamma^0 u^s(p)$. Therefore, the scalar density $n_{s,B}$ in the σ meson field of Eq. (3.41) is given by

$$n_{s,B} = \frac{1}{V} \int d^3x \ \langle F| : \bar{\psi}_B \psi_B : |F\rangle = \frac{2}{(2\pi)^3} \int_0^{p_{F_B}} d^3p \frac{m_B^*}{\sqrt{p^2 + m_B^{*2}}} .$$ (3.192)

The expectation value of σ meson field can be written as,

$$\sigma = \sum_B \frac{g_{\sigma B}}{m_\sigma^2} n_{s,B} .$$ (3.193)

Next we compute the expectation value of the third component of the isovector ρ meson field. In this case, we consider the nucleon doublet. The right hand side of the ρ meson field in Eq. (3.41) is

$$\langle F| : \bar{\psi}_B \gamma^0 \tau_{3B} \psi_B : |F\rangle = \langle F| : \bar{\psi}_p \gamma^0 \tau_{3B} \psi_p : |F\rangle + \langle F| : \bar{\psi}_n \gamma^0 \tau_{3B} \psi_n : |F\rangle$$

$$= \langle F| : \psi_p^\dagger \psi_p : |F\rangle - \langle F| : \psi_n^\dagger \psi_n : |F\rangle .$$

The operator τ_{3B} is the third component of the Pauli spin matrices in the isospin space and operators on the proton and neutron spin states

$$\chi^1 = \begin{pmatrix} 1 \\ 0 \end{pmatrix} \quad ; \quad \chi^2 = \begin{pmatrix} 0 \\ 1 \end{pmatrix} ,$$

giving eigenvalues $\tau_{3B} = +1$ and $\tau_{3B} = -1$, respectively. The mean value of the ρ meson field in Eq. (3.41) is given by

$$m_\rho^2 \rho_{03} = g_{\rho B}(n_p - n_n) .$$ (3.194)

n_p and n_n are proton and neutron densities, respectively. It shows that the ρ meson field plays a very important role in a highly asymmetric matter like the neutron star matter.

The equation of state (EoS) which is a relation between the energy density and pressure, is obtained from the energy-momentum tensor $\Gamma_{\mu\nu}$,

$$\Gamma_{\mu\nu} = -\eta_{\mu\nu}\mathcal{L} + \frac{\partial \mathcal{L}}{\partial(\partial^\mu \varphi)}(\partial_\nu \varphi) .$$ (3.195)

The energy density is explicitly written as,

$$\frac{1}{V} \int d^3x \ \langle : \Gamma_{00} : \rangle = \frac{1}{2}m_\sigma^2\sigma^2 + \frac{1}{3}g_2\sigma^3 + \frac{1}{4}g_3\sigma^4 - \frac{1}{2}m_\omega^2\omega^2 - \frac{1}{2}m_\rho^2\rho_{03}^2$$

$$+ \frac{1}{V} \int d^3x \ \langle : \bar{\psi}_B i\gamma^0 \partial_0 \psi_B : \rangle . \tag{3.196}$$

The last term becomes

$$\frac{1}{V} \int d^3x \ \langle : \bar{\psi}_B i\gamma^0 \partial_0 \psi_B : \rangle = \frac{2}{(2\pi)^3} \int_0^{P_{FB}} d^3p \ E_B ,$$

where $E_B = \sqrt{p^2 + m_B^{*2}} + g_{\omega B}\omega_0 + g_{\rho B}\tau_{3B}\rho_{03}$.
Finally, the total energy density is given by,

$$\frac{1}{V} \int d^3x \ \langle : \Gamma_{00} : \rangle = \frac{1}{2}m_\sigma^2\sigma^2 + \frac{1}{3}g_2\sigma^3 + \frac{1}{4}g_3\sigma^4 + \frac{1}{2}m_\omega^2\omega_0^2 + \frac{1}{2}m_\rho^2\rho_{03}^2$$

$$+ \frac{2}{(2\pi)^3} \sum_B \int_0^{P_{FB}} \sqrt{p^2 + m_B^{*2}} \ d^3p . \tag{3.197}$$

Similarly, the pressure is,

$$\frac{1}{V} \int d^3x \ \langle : \Gamma_{jj} : \rangle = -\frac{1}{2}m_\sigma^2\sigma^2 - \frac{1}{3}g_2\sigma^3 - \frac{1}{4}g_3\sigma^4 + \frac{1}{2}m_\omega^2\omega^2 + \frac{1}{2}m_\rho^2\rho_{03}^2$$

$$+ \frac{1}{V} \int d^3x \ \langle : \bar{\psi}_B i\gamma^j \partial_j \psi_B : \rangle . \tag{3.198}$$

The integrand in the last term can be simplified to,

$$\bar{\psi}_B \gamma^j p_j \psi_B = \psi_B^\dagger (\alpha_j p_j) \psi_B$$

and one has to evaluate $\sum_s u^{\dagger s}(p)(\boldsymbol{\alpha} \cdot \mathbf{p})u^s(p)$. After all these manipulations, the pressure is given by,

$$P = -\frac{1}{2}m_\sigma^2\sigma^2 - \frac{1}{3}g_2\sigma^3 - \frac{1}{4}g_3\sigma^4 + \frac{1}{2}m_\omega^2\omega^2 + \frac{1}{2}m_\rho^2\rho_{03}^2$$

$$+ \frac{1}{3}\frac{2}{(2\pi)^3} \sum_B \int_0^{P_{FB}} \frac{p^2 d^3p}{\sqrt{p^2 + m_B^{*2}}} . \tag{3.199}$$

References

1. Yakovlev, D.G., Haensel, P., Baym, G., Pethick, C.: Lev Landau and the concept of neutron stars. Phys. Usp. **56**(3), 289–295 (2013)
2. Bonolis, L.: Stellar structure and compact objects before 1940: towards relativistic astrophysics. Eur. Phys. J. H **42**(2), 311–393 (2017)
3. Fowler, R.H.: On dense matter. Mon. Not. R. Soc. **87**, 114–122 (1926)
4. Chandrasekhar, S.: The maximum mass of ideal white dwarfs. Astrophys. J. **74**, 81 (1931)
5. Chandrasekhar, S.: On stars, their evolution and their stability. Rev. Mod. Phys. **56**, 137–147 (1984)
6. Chadwick, J.: Possible existence of a neutron. Nature **129**, 312 (1932)
7. Landau, L.D.: To the stars theory. Phys. Zs. Sowjet. **1**, 285 (1932)
8. Baade, W., Zwicky, F.: On super-novae. Proc. Nat. Acad. Sci. **20**(5), 254–259 (1934)
9. Flügge, S.: Der einfluß der neutronen auf den inneren aufbau der sterne. Z. für Astrophys. **6**, 272 (1933)
10. Hund, F.: Materie unter sehr hohen drucken und temperaturen. Ergebn. Exakt. Naturwiss. **15**, 189–228 (1936)
11. Kothari, D.S.: Neutrons, degeneracy and white dwarfs. Proc. R. Soc. **162**(911), 521–528 (1937)
12. Tolman, R.C.: Static solutions of Einstein's field equations for spheres of fluid. Phys. Rev. **55**(4), 364–373 (1939)
13. Oppenheimer, J.R., Volkoff, G.M.: On massive neutron cores. Phys. Rev. **55**(4), 374–381 (1939)
14. Hewish, A.: Pulsars and high density physics. Rev. Mod. Phys. **47**, 567–572 (1975)
15. Hewish, A., Bell, S.J., Pilkington, J.D.H., Scott P.F., Collins R.A.: Observation of a rapidly pulsating radio source. Nature **217**, 709–713 (1968)
16. Gold, T.: Rotating neutron stars as the origin of the pulsating radio sources. Nature **218**, 731–732 (1968)
17. Lattimer, J.M.: The nuclear equation of state and neutron star masses. Annu. Rev. Nucl. Part. Sci. **62**(1), 485–515 (2012)
18. Kramer, M., et al.: The relativistic binary programme on MeerKAT: science objectives and first results. Mon. Not. R. Soc. **504**(2), 2094–2114 (2021)
19. Hu, H., Kramer M., Wex N., Champion D.J., Marcel S.K.: Constraining the dense matter equation-of-state with radio pulsars. Mon. Not. R. Astron. Soc. **497**(3), 3118–3130 (2020)
20. Lyne, A., Graham-Smith, F.: Pulsar Astronomy. Cambridge University Press, Cambridge (2012)
21. Lyne, A.G., et al.: A double-pulsar system: a rare laboratory for relativistic gravity and plasma physics. Science **303**(5661), 1153–1157 (2004)
22. Kramer, M., et al.: Tests of general relativity from timing the double pulsar. Science **314**(5796), 97–102 (2006)
23. Hulse, R.A., Taylor, J.H.: Discovery of a pulsar in a binary system. Astrophys. J. Lett. **195**, L51–L53 (1975)
24. Demorest, P.B., et al.: A two-solar-mass neutron star measured using Shapiro delay. Nature **467**(7319), 1081–1083 (2010)
25. Cromartie, H.T., et al.: Relativistic Shapiro delay measurements of an extremely massive millisecond pulsar. Nat. Astron. **4**, 72–76 (2020)
26. Fonseca, E., et al.: Refined mass and geometric measurements of the high-mass PSR J0740+6620. Astrophys. J. Lett. **915** L12 (2021)
27. Antoniadis, J., et al.: A massive pulsar in a compact relativistic binary. Science **340**(6131), 448 (2013)
28. Romani, R.W., Kandel, D., Filippenko, A.V., Brink, T.G., Zheng, W.: PSR J1810+1744: companion darkening and a precise high neutron star mass. Astrophys. J. **908**(2), L46 (2021)

29. Walter, F.M., Wolk, S.J., Neuhäuser, R.: Discovery of a nearby isolated neutron star. Nature **379**, 233–235 (1996)
30. Walter, F.M., Lattimer J.M.: A revised parallax and its implications for RX J185635–3754. Astrophys. J. **576**(2), L145–L148 (2002)
31. Pons, J.A., et al.: Toward a mass and radius determination of the nearby isolated neutron star RX J185635–3754. Astrophys. J. **564**(2), 981–1006 (2002)
32. Walter, F.M.: RX J1856: astrophysical evidence for quark stars J. Phys. G **30**(1), S461–S470 (2004)
33. Burwitz, V., et al.: The thermal radiation of the isolated neutron star RX–J1856.5–3754 observed with Chandra and XMM-Newton. Astron. Astrophys. **399**, 1109–1114 (2003)
34. Ho, W.C.G., et al.: Magnetic hydrogen atmosphere models and the neutron star RX J1856.5–3754. Mon. Not. R. Astron. Soc. **375**(3), 821–830 (2007)
35. Raaijmakers, G., et al.: A NICER view of PSR J0030+0451: implications for the dense matter equation of state. Astrophys. J. Lett. **887**(1), L22 (2019)
36. Miller, M.C., et al.: PSR J0030+0451 mass and radius from NICER data and implications for the properties of neutron star matter. Astrophys. J. Lett. **887**(1), L24 (2019)
37. Özel, F., Psaltis, D., Arzoumanian, Z., Morsink, S., Baubock: Measuring neutron star radii via pulse profile modeling with NICER. Astrophys. J. **832**(1), 92 (2016)
38. Strohmayer, T.E.: Future probes of the neutron star equation of state using X-ray bursts. AIP Conf. Proc. **714**, 245–252 (2004)
39. Miller, M.C., et al.: The radius of PSR J0740+6620 from NICER and XMM-Newton data. Astrophys. J. Lett. **918**(2), L28 (2021)
40. Riley, T.E., et al.: A NICER view of the massive pulsar PSR J0740+6620 informed by Radio timing and XMM-Newton spectroscopy. Astrophys. J. Lett. **918**(2), L27 (2021)
41. Lattimer, J.M., Schutz, B.F.: Constraining the equation of state with moment of inertia measurements. Astrophys. J. **629**, 979–984 (2005)
42. Damour, T., Schäfer, G.: Higher-order relativistic periastron advances and binary pulsars. Nuo. Cim. B **101**(2), 127–176 (1988)
43. Kramer, M., Wex, N.: The double pulsar system: a unique laboratory for gravity. Clas. Quan. Grav. **26**(7), 073001 (2009)
44. Damour, T., Deruelle, N.: General relativistic celestial mechanics of binary systems. II. The post-Newtonian timing formula. Ann. Inst. Henri Poincaré Phys. Théo **44**(3), 263–292 (1986)
45. Motornenko, A., Steinheimer, J., Stöcker, H.: From cosmic matter to the laboratory. Astron. Nachr. **342**(5), 808–818 (2021)
46. Glendenning, N.K.: The hyperon composition of neutron stars. Phys. Lett. B **114**(6), 392–396 (1982)
47. Glendenning, N.K.: Neutron stars are giant hypernuclei? Astrophys. J. **293**, 470–493 (1985)
48. Chatterjee, D., Vidaña, I.: Do hyperons exist in the interior of neutron stars? Eur. Phys. J. A **52**, 29 (2016)
49. Drago, A., Lavagno, A., Pagliara, G., Pigato, D.: Early appearance of Δ isobars in neutron stars. Phys. Rev. C **90**, 065809 (2014)
50. Migdal, A.B.: Pion fields in nuclear matter. Rev. Mod. Phys. **50**(1), 107–172 (1978)
51. Kaplan, D.B., Nelson, A.E.: Strange goings on in dense nucleonic matter. Phys. Lett. B **175**(1), 57–63 (1986)
52. Nelson, A.E., Kaplan, D.B.: Strange condensate realignment in relativistic heavy ion collisions. Phys. Lett. B **192**(1–2), 193–197 (1987)
53. Glendenning, N.K., Schaffner-Bielich, J.: Kaon condensation and dynamical nucleons in neutron stars. Phys. Rev. Lett. **81**(21), 4564–4567 (1998)
54. Glendenning, N.K., Schaffner-Bielich, J.: First order kaon condensate. Phys. Rev. C **60**(2), 025803 (1999)
55. Brown, G.E., Kubodera, K., Rho, M., Thorsson, V.: A novel mechanism for kaon condensation in neutron star matter. Phys. Lett. B **291**(4), 355–362 (1992)
56. Thorsson, V., Prakash, M., Lattimer, J.M.: Composition, structure and evolution of neutron stars with kaon condensates. Nucl. Phys. A **572**(3–4), 693–731 (1994)

57. Ellis, P.J., Knorren, R., Prakash, M.: Kaon condensation in neutron star matter with hyperons. Phys. Lett. B **349**(1–2), 11–17 (1995)
58. Prakash, M., et al.: Composition and structure of protoneutron stars. Phys. Rep. **280**(1), 1–77 (1997)
59. Muto, T.: Role of weak interaction on kaon condensation in neutron matter: a result with hyperon excitations. Prog. Theo. Phys. **89**(2), 415–435 (1993)
60. Knorren, R., Prakash, M., Ellis, P.J.: Strangeness in hadronic stellar matter. Phys. Rev. C **52**(6), 3470–3482 (1995)
61. Schaffner-Bielich, J., Mishustin, I.N.: Hyperon-rich matter in neutron stars. Phys. Rev. C **53**(3), 1416–1429 (1996)
62. Pons, J.A. et al.: Kaon condensation in proto-neutron star matter. Phys. Rev. C **62**(3), 035803 (2000)
63. Pal, S., Bandyopadhyay, D., Greiner, W.: Antikaon condensation in neutron stars. Nucl. Phys. A **674**(3–4), 553–577 (2000)
64. Banik, S., Bandyopadhyay, D.: Third family of superdense stars in the presence of antikaon condensates. Phys. Rev. C **64**(5), 055805 (2001)
65. Itoh, N.: Hydrostatic equilibrium of hypothetical quark stars. Prog. Theo. Phys. **44**(1), 291–292 (1970)
66. Bodmer, A.R.: Collapsed nuclei. Phys. Rev. D **4**(6), 1601–1606 (1971)
67. Farhi, E., Jaffe, R.L.: Strange matter. Phys. Rev. D **30**(11), 2379–2390 (1984)
68. Witten, E.: Cosmic separation of phases. Phys. Rev. D **30**(2), 272–285 (1985)
69. Rajagopal, K., Wilczek, F.: The condensed matter physics of QCD. In: Shifman, M. (ed.) At the Frontier of Particle Physics: Handbook of QCD, vol. 3, pp 2061–2151. World Scientific, Singapore (2001)
70. Oertel, M., Hempel, M., Klähn, T., Typel, S.: Equations of state for supernovae and compact stars. Rev. Mod. Phys. **89**(1), 015007 (2017)
71. Day, B.D.: Elements of the Brueckner-Goldstone theory of nuclear matter. Rev. Mod. Phys. **39**(4), 719–744 (1967)
72. Rajaraman, R., Bethe, H.A.: Three-body problem in nuclear matter. Rev. Mod. Phys. **39**(4), 745–770 (1967)
73. Brockmann, R., Machleidt, R.: Relativistic nuclear structure. I. Nuclear matter. Phys. Rev. C **42**(5), 1965–1980 (1990)
74. Day, B.D., Wiringa, W.B.: Brueckner-Bethe and variational calculations of nuclear matter. Phys. Rev. C **32**(3), 1057–1062 (1985)
75. Akmal, A., Pandharipande, V.R., Ravenhall, D.G.: Equation of state of nucleon matter and neutron star structure. Phys. Rev. C **58**(3), 1804–1828 (1998)
76. Baldo, M., Burgio, F.: Microscopic theory of the nuclear equation of state and neutron star structure. In: Blaschke, D., Glendenning, N.K., Sedrakian, A. (eds) Lecture Notes in Physics, vol. 578, pp. 1–30. Springer, Heidelberg (2001)
77. Brockmann, R., Machleidt, R.: Nuclear saturation in a relativistic Brueckner-Hartree-Fock approach. Phys. Lett. B **149**(4–5), 283–287 (1984)
78. Horowitz, C.J., Serot, B.D.: Two-nucleon correlations in a relativistic theory of nuclear matter. Phys. Lett. B **137**(5–6), 287–293 (1984)
79. ter Haar, B., Malfliet, R.: Nucleons, mesons and deltas in nuclear matter a relativistic Dirac-Brueckner approach. Phys. Rep. **149**(4), 207–286 (1987)
80. Horowitz, C.J., Serot, B.D.: The relativistic two-nucleon problem in nuclear matter. Nucl. Phys. A **464**(4), 613–699 (1987)
81. Fantoni, S., Fabrocini, A.: Correlated basis function theory for fermion systems. In: Navarro, J., Polls, A. (eds) Microscopic Quantum Many-Body Theories and Their Applications. Lecture Notes in Physics, vol. 510, pp. 119–186. Springer, Berlin (1998)
82. Müther, H., Polls, A.: Two-body correlations in nuclear systems. Prog. Part. Nucl. Phys. **45**(1), 243–334 (2000)
83. Carlson, J., Gandolfi, S., Gezerlis, A., et al.: Quantum Monte Carlo methods for nuclear physics. Rev. Mod. Phys. **87**, 1067–1118 (2015)

84. Akmal, A., Pandharipande, V.R.: Spin-isospin structure and pion condensation in nucleon matter. Phys. Rev. C **56**(4), 2261–2279 (1997)
85. Fantoni, S., Rosati, S.: The hypernetted-chain approximation for a fermion system. Nuo. Cim. A **25**(4), 573–615 (1975)
86. Pandharipande, V.R., Wiringa, R.B.: Variations on a theme of nuclear matter. Rev. Mod. Phys. **51**(4), 821–860 (1979)
87. Coester, F.: Bound states of a many-particle system. Nucl. Phys. A **7**, 421–424 (1958)
88. Coester, F., Kümmel, H.: Short-range correlations in nuclear wave functions. Nucl. Phys. A **16**, 477–485 (1960)
89. Freedman, B.A., MacLerran, L.D.: Fermions and gauge vector mesons at finite temperature and density. I. Formal techniques. Phys. Rev. D **16**(4), 1130–1146 (1977); Fermions and gauge vector mesons at finite temperature and density. II. The ground-state energy of a relativistic electron gas. Phys. Rev. D **16**(4), 1147–1168 (1977); Fermions and gauge vector mesons at finite temperature and density. III. The ground-state energy of a relativistic quark gas. Phys. Rev. D **16**(4), 1169–1185 (1977)
90. Freedman, B.A., MacLerran, L.D.: Quark star phenomenology. Phys. Rev. D **17**(4), 1109–1122 (1978)
91. Fraga, E., Pisarski, R.D., Schaffner-Bielich, J.: Small, dense quark stars from perturbative QCD. Phys. Rev. D **63**(12), 121702 (2001)
92. Drischler, C., Holt, J.W., Wellenhofer, C.: Chiral effective field theory and the high-density nuclear equation of state. Annu. Rev. Nucl. Part. Sci. **71**, 403–432 (2021)
93. Machleidt, R., Sammarruca, F.: Can chiral EFT give us satisfaction? Eur. Phys. J. A **56**(3), 95 (2020)
94. Hebeler, K.: Three-nucleon forces: implementation and applications to atomic nuclei and dense matter. Phys. Rep. **890**, 1–116 (2021)
95. Capano, C.D., et al.: Stringent constraints on neutron-star radii from multimessenger observations and nuclear theory. Nat. Astron. **4**, 625–632 (2020)
96. Skyrme, T.H.R.: The nuclear surface. Phil. Mag. **1**(11), 1043–1054 (1956)
97. Skyrme, T.H.R.: The effective nuclear potential. Nucl. Phys. **9**, 615–634 (1958)
98. Vautherin, D., Brink, D.M.: Hartree-Fock calculations with Skyrme's interaction. Phys. Lett. B **32**(3), 149–153 (1970)
99. Vautherin, D., Brink, D.M.: Hartree-Fock calculations with Skyrme's interaction. I. Spherical nuclei. Phys. Rev. C **5**(3), 626–647 (1972)
100. Stone, J., Reinhard, P.-G.: The Skyrme interaction in finite nuclei and nuclear matter. Prog. Part. Nucl. Phys. **58**(2), 587–657 (2007)
101. Decharge, J., Gogny, D.: Hartree-Fock-Bogolyubov calculations with the D1 effective interaction on spherical nuclei. Phys. Rev. C **21**(4), 1568–1593 (1980)
102. Bandyopadhyay, D., Samanta, C., De, J.N. Samaddar, S.K.: Thermostat properties of finite and infinite nuclear systems. Nucl. Phys. A **511**(1), 1–28 (1990)
103. Banik, S. Bandyopadhyay, D.: Nuclear symmetry energy in the presence of hyperons in the non-relativistic Thomas-Fermi approximation. J. Phys. G **26**(10), 1495–1505 (2000)
104. Walecka, J.D.: A theory of highly condensed matter. Ann. Phys. **83**, 491–529 (1974)
105. Serot, B.D.: A relativistic nuclear field theory with π and ρ mesons. Phys. Lett. B **86**(2), 146–150 (1979)
106. Serot, B.D., Walecka, J.D.: The relativistic nuclear many body problem. Adv. Nucl. Phys. **16**, 1–327 (1986)
107. Serot, B.D., Walecka, J.D.: Recent progress in quantum hadrodynamics. Int. J. Mod. Phys. E **6**(4), 515–631 (1997)
108. Brockman, R., Toki, H.: Relativistic density-dependent Hartree approach for finite nuclei. Phys. Rev. Lett. **68**(23), 3408–3411 (1992)
109. Fuchs, C., Lenske, H., Wolter, H.: Density dependent hadron field theory. Phys. Rev. C **52**(6), 3043–3060 (1995)
110. Lenske, H., Fuchs, C.: Rearrangement in the density dependent relativistic field theory of nuclei. Phys. Lett. B **345**(4), 355–360 (1995)

111. Anastasio, M.R., Celenza, L.S., Pong, W.S., Shakin, C.M.: Relativistic nuclear structure physics. Phys. Rep. **100**(6), 327–392 (1983)
112. Machleidt, R.: The meson theory of nuclear forces and nuclear structure. Adv. Nucl. Phys. **19**, 189–376 (1989)
113. Brockmann, R., Machleidt, R.: Relativistic nuclear structure. I. Nuclear matter. Phys. Rev. C **42**(5), 1965–1980 (1990)
114. Boersma, H.F., Malfliet, R.: From nuclear matter to finite nuclei. I. Parametrization of the Dirac-Brueckner G matrix Phys. Rev. C **49**(1), 233–244 (1994)
115. Huber, H., Weber, F. Weigel, M.K.: Symmetric and asymmetric nuclear matter in the relativistic approach. Phys. Rev. C **51**(4), 1790–1799 (1995)
116. de Jong, F., Lenske, H.: Asymmetric nuclear matter in the relativistic Brueckner-Hartree-Fock approach. Phys. Rev. C **57**(6), 3099–3107 (1998)
117. Müther, H., Prakash, M., Ainsworth, T.L.: The nuclear symmetry energy in relativistic Brueckner-Hartree-Fock calculations. Phys. Lett. B **199**(4), 469–474 (1987)
118. Engvik, L., Hjorth-Jensen, M., Osnes, E., Bao, G., Ostgaard, E.: Asymmetric nuclear matter and neutron star properties. Phys. Rev. Lett. **73**, 2650–2653 (1994)
119. Huber, H., Weber, F., Weigel, M.K., Schaab, C.: Neutron star properties with relativistic equations of state. Int. J. Mod. Phys. E **7**(3), 301–339 (1998)
120. Haddad, S., Weigel, M.: Finite nuclear systems in a relativistic extended Thomas-Fermi approach with density-dependent coupling parameters. Phys. Rev. C **48**(6), 2740–2745 (1993)
121. Cescato, M.L, Ring, P.: Density dependent relativistic mean field theory in deformed nuclei. Phys. Rev. C **57**(1), 134–138 (1998)
122. Typel, S. Wolter, H.H.: Relativistic mean field calculations with density-dependent meson-nucleon coupling. Nucl. Phys. A **656**, 331–364 (1999)
123. Keil, C.M., Hofmann, F., Lenske, H.: Density dependent hadron field theory for hypernuclei. Phys. Rev. C **61**(6), 064309 (2000)
124. Hofmann, F., Keil, C.M., Lenske, H.: Density dependent hadron field theory for asymmetric nuclear matter and exotic nuclei. Phys. Rev. C **64**(3), 034314 (2001)
125. Nikšić, T., Vretenar, D., Finelli, P., Ring, P.: Relativistic Hartree-Bogoliubov model with density-dependent meson-nucleon couplings. Phys. Rev. C **66**(2), 024306 (2002)
126. Hofmann, F., Keil, C.M., Lenske, H.: Application of the density dependent hadron field theory to neutron star matter. Phys. Rev. C **64**(2), 025804 (2001)
127. Nandi, R., Bandyopadhyay, D.: Nuclei in strongly magnetised neutron star crusts. In: Greiner, W. (ed.) Exciting Interdisciplinary Physics, FIAS Interdisciplinary Science Series, pp. 333–343. Springer, Heidelberg (2013)
128. Baym, G., Pethick, C., Sutherland, P.: The ground state of matter at high densities: equation of state and stellar models. Astrophys. J. **170**, 299 (1971)
129. Audi, G., Wapstra, A.H., Thibault, C.: The AME2003 atomic mass evaluation: (II). Tables, graphs and references. Nucl. Phys. A **729**(1), 337–676 (2003)
130. Moller, P., Nix, J.R., Myers, W.D., Swiatecki, W.J.: Nuclear ground-state masses and deformations. At. Data Nucl. Data Tables **59**(2), 185–381 (1995)
131. Langer, W.D., Rosen, L.C., Cohen, J.M., Cameron, A.G.W.: An equation of state at subnuclear densities. Astrophys. Space Sci. **5**(3), 259–271 (1969)
132. Bethe, H.A., Börner, G., Sato, K.: Nuclei in neutron matter. Astron. Astrophys. **7**, 270 (1970)
133. Baym, G., Bethe, H.A., Pethick, C.J.: Neutron star matter. Nucl. Phys. A **175**(2), 225–271 (1971)
134. Ravenhall, D.G., Pethick, C.J., Wilson, J.R.: Structure of matter below nuclear saturation density. Phys. Rev. Lett. **50**(26), 2066–2069 (1983)
135. Oyamatsu, K.: Nuclear shapes in the inner crust of a neutron star. Nucl. Phys. A **561**(3), 431–452 (1993)
136. Cheng, K.S., Yao, C.C., Dai, Z.G.: Properties of nuclei in the inner crusts of neutron stars in the relativistic mean-field theory. Phys. Rev. C **55**(4), 2092–2100 (1997)
137. Negele, J.W., Vautherin, D.: Neutron star matter at sub-nuclear densities. Nucl. Phys. A **207**(2), 298–320 (1973)

138. Bonche, P., Levit, S., Vautherin, D.: Properties of highly excited nuclei. Nucl. Phys. A **427**(2), 278–296 (1984)
139. Bonche, P., Levit, S., Vautherin, D.: Statistical properties and stability of hot nuclei. Nucl. Phys. A **436**(2), 265–293 (1985)
140. Suraud, E.: Semi-classical calculations of hot nuclei. Nucl. Phys. A **462**(1), 109–149 (1987)
141. De, J.N., Vinas, X., Patra, S.K., Centelles, M.: Nuclei beyond the drip line. Phys. Rev. C **64**(5), 057306 (2001)
142. Sil, T. et al., Isospin-rich nuclei in neutron star matter. Phys. Rev. C **66**(4), 045803 (2002)
143. Brack, M., Guet, C., Håkansson, H.B.: Selfconsistent semiclassical description of average nuclear properties' a link between microscopic and macroscopic models. Phys. Rep. **123**(5), 275–364 (1985)
144. Stone, J.R., Miller, J.C., Koncewicz, R., Stevenson, P.D., Strayer, M.R.: Nuclear matter and neutron-star properties calculated with the Skyrme interaction. Phys. Rev. C **68**(3), 034324 (2003)
145. Nandi, R., Bandyopadhyay, D., Mishustin, I.N., Greiner, W.: Inner crusts of neutron stars in strongly quantizing magnetic fields. Astrophys. J. **736**(2), 156 (2011)
146. Hempel, M., Schaffner-Bielich, J.: A statistical model for a complete supernova equation of state. Nucl. Phys. A **837**(3–4), 210–254 (2010)
147. Fái, G., Randrup, J.: Explosion-evaporation model for fragment production in medium-energy nuclear collisions. Nucl. Phys. A **381**(3), 557–576 (1982)
148. Glendenning, N.K.: Compact Stars: Nuclear Physics, Particle Physics and General Relativity. Springer, Heidelberg (1998)
149. Boguta, J., Bodmer, A.R.: Relativistic calculation of nuclear matter and the nuclear surface Nucl. Phys. A **292**(3), 413–428 (1977)
150. Steiner, A.W., Prakash, M., Lattimer, J.M., Ellis, P.J.: Isospin asymmetry in nuclei and neutron stars. Phys. Rep. **411**(6), 325–375 (2005)
151. Steiner, A.W., Hempel, M., Fischer, T.: Core-collapse supernova equations of state based on neutron star observations. Astrophys. J. **774**(1), 17 (2013)
152. Steiner, A.W., Lattimer, J.M., Brown, E.F.: The equation of state from observed masses and radii of neutron stars. Astrophys. J. **722**(1), 33–54 (2010)
153. Sugahara, Y., Toki, H.: Relativistic mean-field theory for unstable nuclei with non-linear σ and ω terms. Nucl. Phys. A **579**(3), 557–572 (1994)
154. Toki, H., Hiratai, D., Sugahara, Y., Sumiyoshi, K., Tanihata, I.: Relativistic many body approach for unstable nuclei and supernova. Nucl. Phys. A **588**, 357–363 (1995)
155. Takahashi, H., et al.: Observation of a $^6_{\Lambda\Lambda}$He double hypernucleus. Phys. Rev. Lett. **87**(21), 212502 (2001)
156. Nakazawa, K., et al.: Double-Λ hypernuclei via the Ξ^- hyperon capture at rest reaction in a hybrid emulsion. Nucl. Phys. A **835**(1), 207–214 (2010)
157. Gal, A., Millener, D.: Shell-model predictions for $\Lambda\Lambda$ hypernuclei. Phys. Lett. B **701**(3), 342–345 (2011)
158. Banik, S., Hempel, M., Bandyopadhyay, D.: New hyperon equations of state for supernovae and neutron stars in density-dependent hadron field theory. Astrophys. J. Suppl. **214**(2), 22 (2014)
159. Weissenborn, S., Chatterjee, D., Schaffner-Bielich, J.: Hyperons and massive neutron stars: vector repulsion and SU(3) symmetry. Phys. Rev. C **85**(6), 065802 (2012); Erratum. Phy. Rev. C **90**, 019904 (2014)
160. Chrien, R.E., Dover, C.B.: Nuclear systems with strangeness. Annu. Rev. Nucl. Part. Sci. **39**(39), 113–150 (1989)
161. Dover, C.B., Gal, A.: Hyperon-nucleus potentials. Prog. Part. Nucl. Phys. **12**, 171–239 (1984)
162. Fukuda, T., et al.: Cascade hypernuclei in the (K^-, K^+) reaction on ^{12}C. Phys. Rev. C **58**, 1306–1309 (1998)
163. Khaustov, P., et al.: Evidence of Ξ hypernuclear production in the $^{12}C(K^-, K^+)^{12}_{\Xi}$Be. Phys. Rev. C **61**(5), 054603 (2000)

164. Schaffner, J., Gal, A.: Properties of strange hadronic matter in bulk and in finite systems. Phys. Rev. C **62**(3), 034311 (2000)
165. Oertel, M., Fantina, A.F., Novak, J.: Extended equation of state for core-collapse simulations. Phys. Rev. C **85**(5), 055806 (2012)
166. Hayano, S., et al.: Evidence for a bound state of the $_\Sigma^4$He hypernucleus. Phys. Lett. B **231**(4), 355–358 (1989)
167. Batty, C.J., Friedman, E., Gal, A.: Density dependence of the Σ nucleus optical potential derived from Σ^- atom data. Phys. Lett. B **335**(3–4), 273–278 (1994)
168. Char, P., Banik, S.: Massive neutron stars with antikaon condensates in a density-dependent hadron field theory. Phys. Rev. C **90**(1), 015801 (2014)
169. Banik, S., Greiner, W., Bandyopadhyay, D.: Critical temperature of antikaon condensation in nuclear matter. Phys. Rev. C **78**(6), 065804 (2008)
170. Batty, C.J., Friedman, E., Gal, A.: Strong interaction physics from hadronic atoms. Phys. Rep. **287**(5), 385–245 (1997)
171. Schaffner, J., et al.: Multiply strange nuclear systems. Ann. Phys. **235**(1), 35–76 (1994)
172. Friedman, E., Gal, A., Mareš, J., Cieplý, A.: K^--nucleus relativistic mean field potentials consistent with kaonic atoms. Phys. Rev. C **60**, 024314 (1999)
173. Koch, V.: K^--proton scattering and the Λ (1405) in dense matter. Phys. Lett. B **337**(3–4), 7–13 (1994)
174. Waas, T., Weise, W.: S-wave interactions of K^- and η mesons in nuclear matter. Nucl. Phys. A **625**(1), 287–306 (1997)
175. Li, G.Q., Lee, C.-H., Brown, G.E.: Kaon production in heavy-ion collisions and maximum mass of neutron stars. Phys. Rev. Lett. **79**, 5124–5217 (1997)
176. Glendenning, N.K., Moszkowski, S.A.: Reconciliation of neutron-star masses and binding of the Λ in hypernuclei. Phys. Rev. Lett. **67**, 2414–2417 (1991)
177. Chodos, A., Jaffe, R.L., Johnson, K., Thom, C.B., Weisskopf, V.F.: New extended model of hadrons. Phys. Rev. D **9**(12), 3471–3495 (1974)
178. Alford, M., Barby, M., Paris, M., Reddy, S.: Hybrid stars that masquerade as neutron stars. Astrophys. J. **629**(2), 969–978 (2005)
179. Weissenborn, S., Sagert, I., Pagliara, G., Hempel, M., Schaffner-Bielich, J.: Quark matter in massive compact stars. Astrophys. J. Lett. **740**(1), L14 (2011)
180. Barrois, B.C.: Superconducting quark matter. Nucl. Phys. B **129**(3), 390–396 (1977)
181. Bailin, D., Love, A.: Superfluidity and superconductivity in relativistic fermion systems. Phys. Rep. **107**(6), 325–385 (1984)
182. Alford, M., Rajagopal, K., Wilczek, F.: QCD at finite baryon density: nucleon droplets and color superconductivity. Phys. Lett. B **422**(1–4), 247–256 (1998)
183. Rapp, R., Schäfer, T., Shuryak, E.V., Velkovsky, M.: Diquark Bose condensates in high density matter and instantons. Phys. Rev. Lett. **81**(1), 53–56 (1998)
184. Alford, M., Rajagopal, K., Wilczek, F.: Color-flavor locking and chiral symmetry breaking in high density QCD. Nucl. Phys. B **537**(1), 443–458 (1999)
185. Alford, M., Rajagopal, K.: Absence of two-flavor color-superconductivity in compact stars. JHEP **06**, 031 (2002)
186. Steiner, A., Reddy, S., Prakash, M.: Color-neutral superconducting quark matter. Phys. Rev. D **66**(9), 094007 (2002)
187. Kaplan, D.B., Reddy, S.: Novel phases and transitions in color flavor locked matter. Phys. Rev. D **65**(5), 054042 (2001)
188. Bedaque, P.F., Schäfer, T.: High-density quark matter under stress. Nucl. Phys. A, **697**, 802–822 (2002)
189. Alford, M., Rajagopal, K., Reddy, S., Wilczek, F.: Minimal color-flavor-locked-nuclear interface. Phys. Rev. D **64**(7), 074017 (2001)
190. Alford, M., Reddy, S.: Compact stars with color superconducting quark matter Phys. Rev. D **67**(7), 074024
191. Nambu, Y., Jona-Lasino, G.: Dynamical model of elementary particles based on an analogy with superconductivity. I. Phys. Rev. **122**(1), 345–358 (1961)

192. Nambu, Y., Jona-Lasino, G.: Dynamical model of elementary particles based on an analogy with superconductivity. II. Phys. Rev. **124**(1), 246–254 (1961)
193. Orsaria, M., Rodrigues, H., Weber, F., Contrera, G.A.: Quark deconfinement in high-mass neutron stars. Phys. Rev. D **89**(1), 015806 (2014)
194. Ranea-Sandoval, I.F., et al.: Constant-sound-speed parametrization for Nambu-Jona-Lasinio models of quark matter in hybrid stars. Phys. Rev. D **93**(4), 045812 (2016)
195. Typel, S., Röpke, G., Klähn, T., Blaschke, D., Wolter, H.H.: Composition and thermodynamics of nuclear matter with light clusters. Phys. Rev. C **81**(1), 015803 (2010)
196. Soma, S., Bandyopadhyay, D.: Properties of binary components and remnant in GW170817 using equations of state in finite temperature field theory models Astrophys. J. **890**(2), 139 (2020)
197. Wick, G.C.: The evaluation of the collision matrix. Phys. Rev. **80**(2), 268–272 (1950)
198. Lattimer, J.M., Lim, Y.: Constraining the symmetry parameters of the nuclear interaction. Astrophys. J. **771**(1), 51 (2013)
199. Tews, I., Lattimer, J.M., Ohnishi, A., Kolometsev, E.E.: Symmetry parameter constraints from a lower bound on neutron-matter energy. Astrophys. J. **848**(2), 105 (2017)
200. Lonardoni, D., Tews, I., Gandolfi, S., Carlson, J.: Nuclear and neutron-star matter from local chiral interactions. Phys. Rev. Res. **2**, 022033 (2020)
201. Hebeler, K., Lattimer, J.M., Pethick, C.J., Schwenk, A.: Equation of state and neutron star properties constrained by nuclear physics and observation. Astrophys. J. **773**(1), 11 (2013)
202. Stone, J.R., Stone, N.J., Moszkowski, S.A.: Incompressibility in finite nuclei and nuclear matter. Phys. Rev. C **89**(4), 044316 (2014)
203. Hornick, N., Tolos, L., Zacchi, A., Christian, J.-E., Schaffner-Bielich, J.: Relativistic parameterizations of neutron matter and implications for neutron stars. Phys. Rev. C **98**(6), 065804 (2018)
204. Agrawal, B., Malik, T., De, J.N., Samaddar, S.K.: Constraining nuclear matter parameters from correlation systematics: a mean-field perspective. Eur. Phys. J. Sp. Top. **230**(2), 517–542 (2021)
205. Wang, M., et al.: The Ame2012 atomic mass evaluation. Chin. Phys. C **36**(12), 1603 (2012)
206. Shlomo, S., Kolomietz, V.M., Coló, G.: Deducing the nuclear-incompressibility coefficient from data on isoscalar compression modes. Eur. Phys. J. A **30**(1), 23–30 (2006)
207. Trippa, L., Colo, G., Vigezzi, E.: Giant dipole resonance as a quantitative constraint on the symmetry energy. Phys. Rev. C **77**(6), 061304 (2008)
208. Brown, B.A.: Neutron radii in nuclei and the neutron equation of state. Phys. Rev. Lett. **85**(25), 5296–5299 (2000)
209. Typel, S., Brown, B.A.: Neutron radii and the neutron equation of state in relativistic models. Phys. Rev. C **64**, 027302 (2001)
210. Centelles, M., Roca-Moza, X., Vinas, X., Warda, M.: Nuclear symmetry energy probed by neutron skin thickness of nuclei. Phys. Rev. Lett. **102**(12), 122502 (2009)
211. Harrison, B.K., Thorne, K.S., Wakano, M., Wheeler, J.A.: Gravitation Theory and Gravitational Collapse. University of Chicago Press, Chicago (1965)
212. Gerlach, U.H.: Equation of state at supranuclear densities and the existence of a third family of superdense stars. Phys. Rev. **172**(5), 1325–1330 (1968)
213. Schaffner-Bielich, J., Hanauske, M., Stöcker, H., Greiner, W.: Phase transition to hyperon matter in neutron stars. Phys. Rev. Lett. **89**(17) 171101 (2002)
214. Glendenning, N.K., Kettner, C.: Possible third family of compact stars more dense than neutron stars. Astron. Astrophys. **353**, L9–L12 (2000)
215. Schertler, K., Greiner, C., Schaffner-Bielich, J., Thoma, M.H.: Quark phases in neutron stars and a third family of compact stars as signature for phase transitions. Nucl. Phys. A **677**(1–4), 463–490 (2000)
216. Banik, S., Bandyopadhyay, D.: Color superconducting quark matter core in the third family of compact stars. Phys. Rev. D **67**(12), 123003 (2003)
217. Banik, S., Hanauske, M., Bandyopadhyay, D., Greiner, W.: Rotating compact stars with exotic matter. Phys. Rev. D **70**(12), 123004 (2005)

218. Alford, M., Han, S., Prakash, M.: Generic conditions for stable hybrid stars. Phys. Rev. D **88**(8), 083013 (2013)
219. Zacchi, A., Tolos, L., Schaffner-Bielich, J.: Twin stars within the SU(3) chiral quark-meson model. Phys. Rev. D **95**(10), 103008 (2017)
220. Christian, J.-E., Schaffner-Bielich, J.: Twin stars and the stiffness of the nuclear equation of state: Ruling out strong phase transitions below 1.7 n_0 with the new NICER radius measurements. Astrophys. J. Lett. **894**(1), L8 (2020)
221. Banik, S., Bandyopadhyay, D.: Bose-Einstein condensation in dense matter and the third family of compact stars. J. Phys. G **28**, 1949–1952 (2002)
222. Glendenning, N.K.: First-order phase transitions with more than one conserved charge: consequences for neutron stars. Phys. Rev. D **46**(4), 1274–1287 (1992)
223. Most, E.R., Weih, L., Rezzolla, L., Schaffner-Bielich, J.: New constraints on radii and tidal deformabilities of neutron stars from GW170817. Phys. Rev. Lett. **120**(26), 261103 (2018)
224. Kettner, C., Weber, F., Weigel, M.K., Glendenning, N.K.: Structure and stability of strange and charm stars at finite temperatures. Phys. Rev. D **51**(4), 1440–1457 (1995)
225. Bardeen, J.M., Thorne, K.S., Meltzer, D.W.: A catalogue of methods for studying the normal modes of radial pulsation of general-relativistic stellar models. Astrophys. J. **145**, 505 (1966)
226. Meltzer, D.N., Thorne, K.S.: Normal modes of radial pulsation of stars at the end point of thermonuclear evolution. Astrophys. J. **145**, 514 (1966)
227. Hessels, J.W.T., et al.: A radio pulsar spinning at 716 Hz. Science **311**, 1901 (2006)
228. Hartle, J.B.: Slowly rotating relativistic stars. I. Equations of structure. Astrophys. J. **150**, 1005 (1967)
229. Hartle, J.B., Thorne, K.S.: Slowly rotating relativistic stars. II. Models for neutron stars and supermassive stars. Astrophys. J. **153**, 807 (1968)
230. Komatsu, H., Eriguchi, Y., Hachisu, I.: Rapidly rotating general relativistic stars. I - Numerical method and its application to uniformly rotating polytropes. Mon. Not. R. Astron. Soc. **237**, 355–379 (1989)
231. Cook, G.B., Shapiro, S.L., Teukolsky, S.A.: Spin-up of a rapidly rotating star by angular momentum loss: effects of general relativity. Astrophys. J. **398**, 203 (1992)
232. Cook, G.B., Shapiro, S.L., Teukolsky, S.A.: Rapidly rotating neutron stars in general relativity: realistic equations of state. Astrophys. J. **424**, 823 (1994)
233. Stergioulas N., Friedman, J.L.: Comparing models of rapidly rotating relativistic stars constructed by two numerical methods. Astrophys. J. **444**, 306 (1995)
234. Nozawa, T., Stergioulas, N., Gourgoulhon, E., Eiguchi, Y.: Construction of highly accurate models of rotating neutron stars - comparison of three different numerical schemes. Astron. Astrophys. Suppl. Ser. **132**, 431–454 (1998)
235. Bonazzola, S., Gourgoulhon, E., Salgado, M., Marck, J.-A.: Axisymmetric rotating relativistic bodies: a new numerical approach for 'exact' solutions. Astron. Astrophys. **278**(2), 421–443 (1993)
236. Salgado, M., Bonazzola, S., Gourgoulhon, E., Haensel, P.: High precision rotating neutron star models. I. Analysis of neutron star properties. Astron. Astrophys. **291**, 155–170 (1994)
237. Marques, M., Oertel, M., Hempel, M., Novak, J.: New temperature dependent hyperonic equation of state: application to rotating neutron star models and I - Q relations. Phys. Rev. C **96**(4), 045806 (2017)
238. Bandyopadhyay, D., Bhat, S., Char, P., Chatterjee, D.: Moment of inertia, quadrupole moment, Love number of neutron star and their relations with strange-matter equations of state. Eur. Phys. J. A **54**, 26 (2018)
239. Gourgoulhon, E., Grandclément, P., Marck, J.-A., Novak, J., Taniguchi, K.: LORENE: spectral methods differential equation solver. In: Astrophysics Source Code Library. http://ascl.net/1608.018
240. Pappas, G., Apostolatos, T.A.: Revising the multipole moments of numerical spacetimes and its consequences. Phys. Rev. Lett. **108**(23), 231104 (2012)
241. Friedman, J.L., Stergioulas, N.: Rotating Relativistic Stars. Cambridge University Press, Cambridge, (2013)

242. Yagi, K., Yunes, N.: I-Love-Q relations in neutron stars and their applications to astrophysics, gravitational waves, and fundamental physics. Phys. Rev. D **88**(2), 023009 (2013)
243. Cipolletta, F., Cherubini, C., Filippi, S., Rueda, J.A., Ruffini, R.: Fast rotating neutron stars with realistic nuclear matter equation of state. Phys. Rev. D **92**(2), (2015) 023007
244. Urbanec, M., Miller, J.C., Stuchlík, Z.: Quadrupole moments of rotating neutron stars and strange stars. Mon. Not. R. Astron. Soc. **433**(3), 1903–1909 (2013)
245. Nandi, R., Bandyopadhyay, D.: Neutron star crust in strong magnetic fields. Magnetised neutron star crusts and torsional shear modes of magnetars. J. Phys. Conf. Ser. **420**(1), 012144 (2013)
246. Kouveluotou, C., et al.: An X-ray pulsar with a superstrong magnetic field in the soft γ-ray repeater SGR1806-20. Nature **393**, 235–237 (1998)
247. Kouveluotou, C., et al.: Discovery of a magnetar associated with the soft gamma repeater SGR 1900+14. Astrophys. J. **510**(2), L115–L118 (1999)
248. Thompson, C., Duncan, R.C.: Neutron star dynamos and the origins of pulsar magnetism. Astrophys. J. **408**, 194 (1993)
249. Thompson, C., Duncan, R.C.: The soft gamma repeaters as very strongly magnetized neutron stars. II. Quiescent neutrino, x-ray, and Alfven wave emission. Astrophys. J.**473**, 322 (1996)
250. Chandrasekhar, S., Fermi, E.: Problems of gravitational stability in the presence of a magnetic field. Astrophys. J. **118**, 116 (1953)
251. Lai, D., Shapiro, S.L.: Cold equation of state in a strong magnetic field: effects of inverse β-decay. Astrophys. J. **383**, 745 (1991)
252. Nandi, R., Bandyopadhyay, D.: Neutron star crust in strong magnetic fields. J. Phys. Conf. Scr. **312**(4), 042016 (2011)
253. Nandi, R., Char, P., Chatterjee, D., Bandyopadhyay, D.: Role of nuclear physics in oscillations of magnetars. Phy. Rev. C **94**(2), 025801 (2016)
254. Chakrabarty, S., Bandyopadhyay, D., Pal, S.: Dense nuclear matter in a strong magnetic field. Phys. Rev. Lett. **78**(15), 2898–2901 (1997)
255. Kobayashi, M., Sakamoto, M.: Radiative corrections in a strong magnetic field. Prog. Theor. Phys. **70**(5), 1375–1384 (1983)
256. Broderick, A.E., Prakash, M., Lattimer, J.M.: The equation of state of neutron star matter in strong magnetic fields. Astrophys. J. **537**(1), 351–367 (2000)
257. Horowitz, C.J., Serot, B.D.: Self-consistent Hartree description of finite nuclei in a relativistic quantum field theory. Nucl. Phys. A **368**(3), 503–528 (1981)
258. Lattimer, J.M., Pethick, C.J., Prakash, M., Haensel, P.: Direct URCA process in neutron stars. Phys. Rev. Lett. **66**(21), 2701–22704 (1991)
259. Pethick, C.J.: Cooling of neutron stars. Rev. Mod. Phys. **64**, 1133–1140 (1992)
260. Bandyopadhyay, D., Chakrabarty, S., Pal, S.: Quantizing magnetic field and quark-hadron phase transition in a neutron star. Phys. Rev. Lett **79**(12), 2176–2179 (1997)
261. Chatterjee, D., Novak, J., Oertel, M.: Magnetic field distribution in magnetars. Phys. Rev. C **99**(5), 055811 (2019)
262. Lai, D.: Matter in strong magnetic fields. Rev. Mod. Phys. **73**(3), 629 (2001)
263. Bandyopadhyay, D., Chakrabarty, S., Pal, S., Dey, P.: Rapid cooling of magnetized neutron stars. Phys. Rev. D **58**(12), 121301 (R) (1998)
264. Baiko, D.A., Yakovlev, D.G.: Direct URCA process in strong magnetic fields and neutron star cooling. Astron. Astrophys. **342**, 192–200 (1999)
265. Broderick, A.E., Prakash, M., Lattimer, J.M.: Effects of strong magnetic fields in strange baryonic matter. Phys. Lett. B **531**(3–4), 167–174 (2002)
266. Sinha, M., Bandyopadhyay, D.: Hyperon bulk viscosity in strong magnetic fields. Phys. Rev. D **79**(12), 123001 (2009)
267. Dey, P., Bhattacharyya, A., Bandyopadhyay, D.: Bose-Einstein condensation in dense nuclear matter and strong magnetic fields. J. Phys. G **28**(8), 2179–2186 (2002)
268. Baiotti, L., Rezzolla, L.: Binary neutron star mergers: a review of Einstein's richest laboratory. Rep. Prog. Phys. **80**(9), 096901 (2017)

269. Hillebrandt, W., Wolff, R.G.: Models of type II supernova explosions. In: Arnett, W.D., Truran, J.M. (eds.) Nucleosynthesis: Challenges and New Developments, p. 131. University of Chicago Press, Chicago (1985)
270. Lattimer, J.M., Swesty, F.D.: A generalized equation of state for hot, dense matter. Nucl. Phys. A **535**(2), 331–376 (1991)
271. Shen, H., Toki, H., Oyamatsu, K., Sumiyoshi, K.: Relativistic equation of state of nuclear matter for supernova and neutron star. Nucl. Phys. A **637**(3), 435–450 (1998) 435
272. Raduta, Ad.R., Gulminelli, F.: Statistical description of complex nuclear phases in supernovae and proto-neutron stars. Phys. Rev. C **82**(6), 065801 (2010)
273. Shen, G., Horowitz, C.J., Teige, S.: Equation of state of nuclear matter in a virial expansion of nucleons and nuclei. Phys. Rev. C **82**(4), 045802 (2010)
274. Shen, G., Horowitz, C.J., Teige, S.: New equation of state for astrophysical simulations. Phys. Rev. C **83**(3), 035802 (2011)
275. Shen, G., Horowitz, C.J., O'Connor, E.: Second relativistic mean field and virial equation of state for astrophysical simulations. Phys. Rev. C **83**(6), 065808 (2011)
276. Fischer, T., et al.: Core-collapse supernova explosions triggered by a quark-hadron phase transition during the early post-bounce phase. Astrophys. J. Suppl. Ser. **194**(2), 39 (2011)
277. Blinnikov, S.I., Panov, I.V., Rudzsky, M.A., Sumiyoshi, K.: The equation of state and composition of hot, dense matter in core-collapse supernovae. Astron. Astrophys. **535**, A37 (2011)
278. Hempel, M., Fischer, T., Schaffner-Bielich, J., Liebendörfer, L.: New equations of state in simulations of core-collapse supernovae. Astrophys. J. **748**(1), 70 (2012)
279. Buyukcizmeci, N., Botvina, A.S., Mishustin, I.N.: Tabulated equation of state for supernova matter including full nuclear ensemble. Astrophys. J. **789**(1), 33 (2014)
280. Togashi, H., Takano, H., Sumiyoshi, K., Nakazato, K.: Application of the nuclear equation of state obtained by the variational method to core-collapse supernovae. Prog. Theor. Exp. Phys. **2014**(2), 023D05 (2014)
281. Ishizuka, C., Ohnishi, A., Tsubakihara, K., Sumiyoshi, K., Yamada, S.: Tables of hyperonic matter equation of state for core-collapse supernovae. J. Phys. G **35**(8), 085201 (2008)
282. Nakazato, K., Sumiyoshi, K., Yamada, S.: Astrophysical implications of equation of state for hadron-quark mixed phase: compact stars and stellar collapses. Phys. Rev. D **77**(10), 103006 (2008)
283. Sagert, I., Fischer, T., Hempel, M., Pagliara, G., Schaffner-Bielich, J.: Signals of the QCD phase transition in core-collapse supernovae. Phys. Rev. Lett. **102**(8), 081101 (2009)
284. Sumiyoshi, K., Ishizuka, C., Ohnishi, A., Yamada, S., Suzuki, H.: Emergence of hyperons in failed supernovae: trigger of the black hole formation. Astrophys. J. Lett. **690**(1), L43–L46 (2009)
285. Shen, H., Toki, H., Oyamatsu, K., Sumiyoshi, K.: Relativistic equation of state for core-collapse supernova simulations. Astrophys. J. Suppl. Ser. **197**(2), 20 (2011)
286. Nakazato, K., et al.: Hyperon matter and black hole formation in failed supernovae. Astrophys. J. **745**(2), 197 (2012)
287. Peres, B., Oertel, M., Novak, J.: Influence of pions and hyperons on stellar black hole formation. Phys. Rev. D **87**(4), 043006 (2013)
288. Fischer, T., Hempel, M., Sagert, I., Suwa, Y., Schaffner-Bielich, J.: Symmetry energy impact in simulations of core-collapse supernovae. Eur. Phys. J. A **50**, 46 (2014)
289. Malik, T., Bandyopadhyay, D., Banik, S.: New equation of state involving Bose-Einstein condensate of antikaon for supernova and neutron star merger simulations. Eur. Phys. J. Sp. Top. **230**(2), 561–566 (2021)
290. Malik, T., Bandyopadhyay, D., Banik, S.: Equation-of-state table with hyperon and antikaon for supernova and neutron star merger. Astrophys. J. **910**(2), 96 (2021)
291. Bandyopadhyay, D.: Neutron stars: laboratories for fundamental physics under extreme astrophysical conditions. J. Astrophys. Astron. **38**(3), 37 (2017)
292. Brown, G.E., Bethe, H.A.: A scenario for a large number of low-mass black holes in the galaxy. Astrophys. J. **423** 659 (1994)

293. Baumgarte, T.W., Janka, H.-Th., Keil, W., Shapiro, S.L., Teukolsky, S.A.: Delayed collapse of hot neutron stars to black holes via hadronic phase transitions. Astrophys. J. **468**, 823 (1996)
294. Prakash, M., Cooke, J.R., Lattimer, J.M.: Quark-hadron phase transition in protoneutron stars. Phys. Rev. D **52**(2), 661–665 (1995)
295. Banik, S.: Probing the metastability of a protoneutron star with hyperons in a core-collapse supernova. Phys. Rev. C **89**(3), 035807 (2014)
296. Keil, W., Janka, H.-Th.: Hadronic phase transitions at supranuclear densities and the delayed collapse of newly formed neutron stars. Astrophys. J. **296**, 145 (1995)
297. O'Connor, E., Ott, C.D.: Black hole formation in failing core-collapse supernovae. Astrophys. J. **730**, 70 (2011)
298. Char, P., Banik, S., Bandyopadhyay, D.: A comparative study of hyperon equation of state in supernova simulations. Astrophys. J. **809**(2), 116 (2015)

Chapter 4
Binary Neutron Star Mergers

Summary Gravitational waves (GWs) from neutron stars were long thought to be important probes of dense matter in its interior. Albert Einstein predicted the existence of GWs in his theory of general relativity more than a century back. The discoveries of compact astrophysical objects in various astrophysical events such as neutron stars in supernova explosions, double neutron star systems were made using radio, x-ray, and optical telescopes over the past fifty years or so. This progress in astronomy enthused astronomers to build gravitational wave detectors. The direct detection of GWs was confirmed when the first binary black hole merger was recorded by the GW detectors on the earth in 2015. This was followed by the first discovery of a binary neutron star (BNS) merger in 2016. After that, GWs from large numbers of binary compact object mergers were detected. It was possible to extract the tidal deformability from the late inspiral of the BNS merger that determined the radii of neutron stars. An upper bound on the maximum mass of neutron stars was estimated from the remnant of the first BNS merger. We discuss the determination of the properties of neutron stars and the dense matter in it, using GW observations in this chapter.

4.1 Gravitational Waves as New Window into Neutron Stars

The theoretical prediction of gravitational waves was made more than a century ago. Einstein worked on the relativistic theory of gravity and published it in 1915 [1]. He observed that his theory could predict gravitational waves [2]. However, Einstein was not so sure about his theory producing GWs. He and his collaborator obtained non-linear solutions of the field equations leading to gravitational waves. But they could not find nonsingular solutions to this problem and sent the manuscript to the Physical Review concluding that GWs did not exist. However, Einstein corrected this mistake later and it was published in a not so renowned journal later [3].

© The Author(s), under exclusive license to Springer Nature Switzerland AG 2022   135
D. Bandyopadhyay, K. Kar, *Supernovae, Neutron Star Physics and Nucleosynthesis*, Astronomy and Astrophysics Library,
https://doi.org/10.1007/978-3-030-95171-9_4

GWs are ripples in the fabric of spacetime due to accelerating masses. The gravitational wave equation is obtained by writing the spacetime metric,

$$g_{\mu\nu} = \eta_{\mu\nu} + h_{\mu\nu} \,, \tag{4.1}$$

and linearizing Einstein field equations. Here $\eta_{\mu\nu}$ and $h_{\mu\nu}$ are the flat metric and perturbation, respectively. Assuming a small perturbation, higher order terms of the perturbation and their derivatives are neglected and a wave equation of $h_{\mu\nu}$ describing GWs is obtained.

The direct detection of gravitational waves was possible a century later. But many interesting astrophysical events happened during this period leading to the discovery of GWs. One such important discovery was the first detection of a pulsar by Jocelyn Bell and Antony Hewish in 1967 [4]. The first binary pulsar PSR1913+16 was discovered by R. A. Hulse and J. H. Taylor in 1974 [5]. The Hulse–Taylor binary pulsar provided the first evidence of indirect detection of gravitational waves. The pulsar in the Hulse–Taylor binary is being monitored since its discovery and the estimate of the cumulative decay of the orbital period ($\dot{P}_b$) of the Hulse–Taylor pulsar due to gravitational radiation with time matches with the prediction of Einstein's theory of general relativity. It was also possible to determine that neutron stars in the Hulse–Taylor binary would collide in $\sim$300 million years [6]. All these developments prepared the ground for building gravitational wave detectors to record tiny signals created in cataclysmic astrophysical events like binary compact object coalescences from the earth.

Joseph Weber started the field of gravitational wave detection and built gravitational wave detectors based on the idea of resonant bar detectors that might vibrate during the passage of gravitational waves [7, 8]. But his efforts did not produce any discovery of GWs [9]. This was followed by the construction of GW detectors using similar basic principles of Michelson interferometer. This led to the construction of the Laser Interferometer Gravitational-wave Observatory (LIGO) detectors at Livingstone and Hanford, USA, very wide band interferometric gravitational wave antenna (Virgo) detector near Pisa, Italy and German/British gravitational wave detector GEO600 in Hannover, Germany [10–12]. Interferometers are to be large in dimension for detecting GWs. These detectors are operational since 2005. Another detector named KAGRA in Japan joined this network of gravitational wave detectors in 2020. The activity of LIGO is expanded to India for building LIGO, India which would be commissioned in the latter part of this decade. Other new proposals of big detectors are Einstein telescope, Cosmic Explorer and a space based gravitational wave detector known as the Laser Interferometer Space Antenna (LISA). LIGO, Virgo, and other detectors are sensitive to 10–100 Hz GWs whereas the LISA would probe GWs in the mHz regime. Radio astronomers are using a network of very stable millisecond pulsars and looking for tiny changes in arrival times of pulses due to the propagation of nano Hertz gravitational waves emitted by massive black holes of $\sim 10^9$ $M_\odot$. This set-up is known as the pulsar timing array (PTA) [13]. Radio astronomers are involved in several PTA projects around the world [13].

The first discovery of binary black hole (BBH) merger GW150914 by the advanced LIGO detectors was the first direct proof of the presence of black holes in the universe [14]. More surprise was in store when GW signals of the first BNS merger GW170817 was detected by the advanced LIGO-Virgo collaboration (LVC) [15]. The inspiral signal of GW170817 lasted for 100 s. This was quite a loud signal having a signal to noise ratio of 32.4. The LIGO and Virgo network of detectors together pin-pointed the sky position localization to 28 deg^2 that led to a very successful electromagnetic campaign to search for a transient [15–17]. The gravitational wave source and the associated short Gamma Ray Bursts (sGRB) were the nearest ever sources with a luminosity distance of 40^{+8}_{-14} Mpc [15]. The BNS merger GW170817 was not only discovered in GWs, it was also seen in light. This heralded a new beginning of the multimessenger astrophysics along with GWs and opened up a new avenue to investigate the dense matter in neutron stars [16]. This was followed by the discovery of the BNS merger GW190425 [18] and black hole—neutron star mergers (GW190814, GW200105, and GW200115) by the LIGO/Virgo/KAGRA [19, 20]. However, no EM counterparts were observed in those events. Furthermore, a large number of BBH mergers were recorded so far by the GW detectors.[1]

It is now realized that the discoveries of GWs from coalescing compact objects have the same potential as Galileo's observations with an optical telescope. Similarly, the impact of optical, x-ray, and radio telescopes in astronomy was enormous before the advent of GW detectors. The discoveries of GW signals from binary compact object coalescences have started to reveal the explosive phenomena in the universe and the interesting astrophysics associated with those.

4.2 First Binary Neutron Star Merger GW170817 and Multimessenger Astrophysics

The first binary neutron star merger event GW170817 discovered by the LVC was detected in gravitational waves and EM radiations. GW signals were followed by a sGRB 1.7 s after the merger [17]. This confirmed the link between a neutron star merger event and sGRB. EM signals from the transient were observed in ultraviolet (UV), optical and infrared (IR) bands. This also proved the prediction of a kilonova associated with the BNS merger. The kilonova was powered by the decays of radioactive r-process nuclei synthesized in the ejected matter in the BNS merger [21]. This heralded a new era of the multimessenger astrophysics along with GWs.

The discovery of GW170817 provided a wealth of information about the sGRB, binary chirp mass, tidal deformability, speed of gravitational waves and Hubble constant [15–17]. However, no GW signal from the remnant was observed because the LIGO detectors are not sensitive to kilo Hertz GWs [22]. This event rejuvenated

---

[1] https://www.nature.com/articles/d41586-021-03089-y.

nuclear astrophysics communities around the world because compositions and dense matter equation of state (EoS) in neutron stars and r-process nucleosynthesis in the ejected neutron-rich matter could be probed [21, 23]. The EM counterpart following the merger led to important clues to the fate of the remnant. The chirp mass was determined to be $1.186^{+0.001}_{-0.001}$ $M_\odot$ [24]. Neutron star masses in the binary were estimated to be in the range 1.16–1.60 $M_\odot$ for low spins supported by the observations of the galactic neutron stars whereas the mass range was 1.00–1.89 $M_\odot$ for high spins. The massive remnant formed in the merger had a mass of $2.73^{+0.04}_{-0.01}$ and $2.77^{+0.22}_{-0.05}M_\odot$ for the low and high spin cases, respectively, in the 90% credible intervals [24].

The EM counterpart of GW170817 indicated that a short-lived hypermassive neutron star (HMNS) was most likely born in this event. The collapse of the HMNS to a black hole could set an upper limit on the maximum mass of non-rotating neutron stars [23, 25–28]. On the other hand, the lower bound on the maximum mass of neutron stars (2.08 ± 0.07 $M_\odot$) is obtained from the galactic pulsar observations [29]. The upper and lower bounds on the maximum masses of neutron stars together result in strong constraints on the EoS of dense matter [25, 30, 31]. The tidal deformability parameter was extracted from the late inspiral gravitational wave data of GW170817 [14]. Consequently, radii of neutron stars in the binary were estimated from the combined tidal deformability [32–34]. This facilitated the determination of the EoS of neutron star matter.

4.3 Tidal Deformability, Love Number, and EoS

In the inspiral stage of a BNS, one neutron star is deformed due to the tidal effects of its companion. Here we are interested in the quadrupolar tidal field. In the rest frame of the neutron star, the quadrupolar tidal field (E_{ij}) arising due to the distortion of the spacetime by the companion is $E_{ij} = R_{titj}$, where R_{titj} is the component of the Riemann tensor $R_{\mu\alpha\nu\beta}$ [35–38]. Consequently the tidally deformed neutron star would influence the exterior spacetime that are recognized through multipole moments at asymptotically large distances. The expansion of the time-time component (g_{tt}) of the metric in the local asymptotic rest frame of the neutron star, also known as asymptotically Cartesian coordinate system with the origin at the center of mass of the neutron star is given by [35–38]

$$\mathcal{L}t_{r\to\infty} \frac{(1+g_{tt})c^2}{2} = \frac{GM}{r} + \frac{(3n^i n^j - \delta^{ij})Q_{ij}}{2r^3} + O\left(\frac{1}{r^4}\right) - \frac{1}{2}n^i n^j E_{ij}r^2 + O(r^3),$$

$$(4.2)$$

where n^i and n^j denote unit vectors. Here the first term implies the monopole term, the second term corresponds to the quadrupole term with the mass quadrupole moment tensor Q_{ij}. The dipole term is not present because the origin of the coordinate system coincides with the center of mass of the neutron star. Both E_{ij} and Q_{ij} are symmetric and traceless.

The quadrupolar tidal field induces a quadrupole moment in a neutron star. The phase evolution of GW signals contains information on the tidal deformability during the later part of the inspiral. In the case of a static spherically symmetric star placed under the influence of a static external quadrupolar tidal field E_{ij}, the induced quadrupole moment (Q_{ij}) to the linear order is given by

$$Q_{ij} = -\lambda E_{ij} , \tag{4.3}$$

where the coefficient λ is the tidal deformability. The following points are considered in arriving at the relation of Eq. (4.3). The tidal field excites oscillation modes of the neutron star labeled by a set of integers n, m, ℓ, where n denotes the radial nodes and (ℓ, m) correspond to the angular quantum numbers of a spherical harmonic decomposition. The tidally induced quadrupole moment is a sum of all $\ell = 2$ oscillation modes for different radial nodes. The fundamental modes with $n = 0$ would dominate the sum and the contributions of other radial nodes in the sum might be neglected [38]. If the characteristic time scales of the fundamental modes are shorter than the time scale of variations in the tidal field, a linear relation in Eq. (4.3) might be justified.

A dimensionless quantity, k_2, known as the Love number is defined as

$$k_2 = \frac{3}{2} \frac{\lambda}{R^5} . \tag{4.4}$$

Here k_2 is obtained from the leading order for tidal effects, $\ell = 2$.

Tanja Hinderer and collaborators determined the quadrupole moment in a static external electric tidal field that were extracted as coefficients from an asymptotic expansion of the metric at large distances [37, 39, 40]. The quadrupole deformation is produced by an electric-type (or even parity) tidal perturbation. If the deformation is small, the solutions of tidally deformed neutron stars could be constructed in the same way as slowly rotating neutron stars in the Hartle-Thorne formalism as discussed in Chap. 3. The linear quadrupolar perturbation is decomposed into spherical harmonics in this problem. Further $m = 0$ is considered in this calculation [37]. Applying linear $\ell = 2$ static, even parity perturbations onto the spherically symmetric star and working in the Regge-Wheeler gauge, the deformed metric is given by [40–42],

$$ds^2 = -e^{2\nu(r)}\left[1 + H(r)Y_{20}(\theta, \phi)\right]dt^2 + e^{2\gamma(r)}\left[1 - H(r)Y_{20}(\theta, \phi)\right]dr^2$$
$$+ r^2\left[1 - K(r)Y_{20}(\theta, \phi)\right](d\theta^2 + \sin^2\theta d\phi^2) . \tag{4.5}$$

The relation between $H(r)$ and $K(r)$ is obtained as,

$$\frac{dK(r)}{dr} = \frac{dH(r)}{dr} + 2H(r)\nu(r) . \tag{4.6}$$

For the perturbation to the time-time component of the metric, the perturbed Einstein field equations are combined to get the following second order differential equation of $H(r)$ [37, 40],

$$\frac{d^2 H}{dr^2} + \left(\frac{2}{r} + \frac{dv}{dr} - \frac{d\gamma}{dr} \right) \frac{dH}{dr} + \left[-\frac{6e^{2\gamma}}{r^2} - 2 \left(\frac{dv}{dr} \right)^2 + 2 \frac{d^2 v}{dr^2} + \frac{3}{r} \frac{d\gamma}{dr} + \frac{7}{r} \frac{dv}{dr} \right.$$

$$\left. -2 \frac{dv}{dr} \frac{d\gamma}{dr} + \frac{f}{r} \left(\frac{dv}{dr} + \frac{d\gamma}{dr} \right) \right] H = 0 , \tag{4.7}$$

where the EoS function $f = d\varepsilon/dp$.

The second order differential equation of $H(r)$ in Eq. (4.7) is decomposed into a set of first order ordinary differential equations introducing $\alpha = \frac{dH}{dr}$ [37, 38, 40]. These differential equations are solved simultaneously along with the TOV and the associated equations of Sect. 3.6 in Chap. 3 starting from the center up to the surface of the star. This gives the interior solution of the problem.

The general solution of the differential equation for H outside the star is written as [37, 38],

$$H = a_1 Q_{\ell 2} \left(\frac{r}{M} - 1 \right) + a_2 P_{\ell 2} \left(\frac{r}{M} - 1 \right) , \tag{4.8}$$

where $P_{\ell 2}$ and $Q_{\ell 2}$ are the associated Legendre functions and a_1 and a_2 are unknown coefficients. For the determination of a_1, and a_2, the asymptotic solution of Eq. (4.8) at $r \to \infty$ is compared with Eq. (4.2) and the unknown coefficients of the asymptotic solution of Eq. (4.8) are connected with the quadrupole moment and tidal field of Eq. (4.2). Finally, using Eq. (4.3), the coefficients are given by [37]

$$a_1 = \frac{15}{8M^3} \lambda E, \quad a_2 = \frac{1}{3} M^2 E , \tag{4.9}$$

where E is the magnitude of the tidal field tensor.

The interior and exterior solutions and their derivatives are matched at the surface of the star to get the values of λ. Consequently the $l = 2$ Love number, k_2, is obtained using Eq. (4.4) [37]

$$k_2 = \frac{8C^5}{5} (1 - 2C)^2 [2 + 2C(y - 1) - y]$$

$$\times \{ 2C[6 - 3y + 3C(5y - 8)] + 4C^3 [13 - 11y + C(3y - 2) + 2C^2(1 + y)]$$

$$+ 3(1 - 2C)^2 [2 - y + 2C(y - 1)] \ln(1 - 2C) \}^{-1} , \tag{4.10}$$

where the compactness parameter of a neutron star $C = \frac{M}{R}$ and

$$y = \frac{R\alpha(R)}{H(R)} . \tag{4.11}$$

However, the term $(\frac{f}{r})(\frac{dv}{dr} + \frac{dy}{dr})$ in Eq. (4.7) becomes infinitely large at the surface of a quark star or a compressible star due to a nonzero energy density [40, 44]. As a result, $\frac{dH}{dr}$ is discontinuous at the surface. This demands a modification of Eq. (4.11) as implemented in [40]

$$y = \frac{R\alpha(R)}{H(R)} - \frac{4\pi R^3 \varepsilon_{in}}{M},$$
(4.12)

where ε_{in} is the energy density just below the surface.

The Love number and tidal deformability of neutron stars are sensitive to the compositions and EoS of neutron star matter. Equation (4.7) along with TOV equations is solved supplying the EoS as an input. Here we adopt EoSs with different compositions of Table 3.1 denoted by the DD2, BHBΛϕ, hybrid and described in details in Sect. 3.6. All these EoSs are constructed in the density dependent hadron field (DDRH) theoretical model and the hybrid EoS implies a first order phase transition from hadronic to quark matter. In this calculation, we adopt another EoS involving Λ hyperons and the Bose–Einstein condensate of K^- mesons that was computed in the DDRH model. This EoS is denoted as BHBΛ$K^-\phi$ and obtained for an antikaon optical potential depth of -120 MeV [43, 45, 46]. The $\ell = 2$ love number (k_2) is exhibited with the compactness ($C = M/R$) in Fig. 4.1 for above mentioned EoSs. The Love number is an indicator of how easy or difficult it would be to deform a neutron star. For all EoSs, we note that the Love number decreases with the increasing compactness. This shows that a more compact star has a smaller

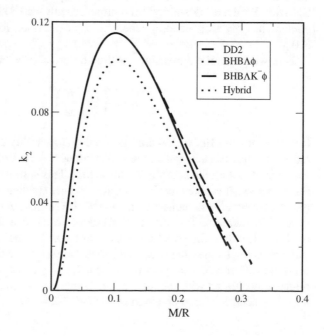

Fig. 4.1 The quadrupolar love number is plotted with the compactness of neutron stars for EoSs with different compositions. This is taken from Ref. [43] and reproduced with kind permission of the European Physical Journal

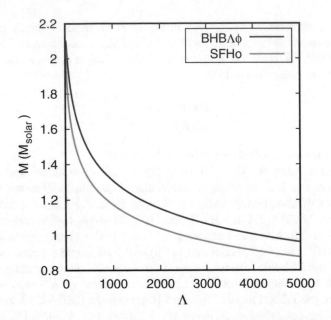

Fig. 4.2 The dimensionless tidal deformability is shown as a function of neutron star mass for BHB$\Lambda\phi$ and SFHo EoSs. This is taken from Ref. [31] and reproduced by permission of the IoP Publishing

Love number. Furthermore, this effect is more pronounced in softer EoSs like BHB$\Lambda\phi$, BHB$\Lambda K^-\phi$ and hybrid compared with the DD2 made of nucleons-only matter. The Love number approaches zero for black holes for which the compactness is 0.5.

Now we define the dimensionless tidal deformability (Λ) using Eq. (4.4) as

$$\Lambda = \frac{2}{3}k_2\left(\frac{R}{M}\right)^5.$$

(4.13)

It is evident from Eq. (4.13) that the tidal deformability depends on the mass and radius of a neutron star which are direct probes of the EoS. Figure 4.2 depicts the variation of Λ with the neutron star mass [31]. This is shown for a soft and stiff EoS. For both EoSs, it is noted that heavier masses have smaller tidal deformabilities. This might have observable consequence in the determination of the tidal deformability from BNS mergers. The stiff EoS BHB$\Lambda\phi$ leads to a higher value of the tidal deformability than the soft EoS SFHo for a fixed value of the mass of a neutron star. For example, the tidal deformabilities corresponding to the BHB$\Lambda\phi$ and SFHo EoSs are 697 and 334, respectively, for a 1.4 M$_\odot$ neutron star [31]. These values might be compared with the 90% upper bound on the tidal deformability ($\Lambda_{1.4}$) of a 1.4 M$_\odot$ neutron star as obtained from GW170817, i.e. $\Lambda_{1.4} < 580$ [32].

4.4 I-Love-Q Universal Relations

The knowledge of the EoS at high densities is not complete. All efforts are being made to constrain the EoS in the high density regime from various astrophysical observations that we have already described in Chap. 3 and are going to be discussed in the following sections. In these circumstances, EoS insensitive or universal relations are playing important roles in explaining astrophysical observations involving neutron stars, for example, BNS mergers. The description of the exterior spacetime of a compact object independent of its compositions and EoS is the main objective to study universal relations among various observables of compact stars. Such universal relations involving the momentum of inertia (I), quadrupole momentum (Q) and Love number or tidal deformability popularly known as I-Love-Q relations were first explored by Yagi and Yunes [47, 48]. It was noted that the results of different EoSs for those observables deviated by a small amount from the universality. This kind of relation was also shown to exist for other observables of neutron stars [49–52]. For example, the relation between the ratio of the maximum mass at the mass-shedding limit and the maximum mass of a static neutron star with the normalized angular momentum with respect to the maximum angular momentum shows an approximate universal behavior [49–51]. However, it is noted that universal relations might not be valid under certain physical conditions. It was found that the universal relations at high rotational frequencies ~1 kHz and high magnetic field ~10^{13} G were not satisfied [53, 54]. The violation of I-Love-Q relations was reported in case of dense matter undergoing a first order phase transition from hadronic to quark matter in neutron stars [43].

The tidal deformability parameter, the moment of inertia, and the quadrupole moment are sensitive to the EoS. However, K.Yagi and N. Yunes demonstrated that the relationship involving any two of I, Q and Λ was EoS insensitive for slowly rotating neutron stars [47, 48]. Later, it was shown that the approximate universal relation was found to hold good between I and Q even for rapidly rotating neutron stars [55]. These approximate universal relations have immense physical implications. For example, the measurement of the quadrupole moment of a neutron star is quite difficult. On the other hand, it would be possible to estimate the moment of inertia of pulsar A in the double pulsar system PSR J0737-3039 [56]. An approximate universal $I-Q$ relation might be used for the determination of the quadrupole of the neutron star. Furthermore, similar approximate universal relations of Λ with I or Q could be used to compute the moment of inertia and quadrupole moment of a binary component because the tidal deformability was extracted from the gravitational wave signal of GW170817 [31].

We discuss the approximate universal relations among the dimensionless moment of inertia ($\bar{I} = I/M^3$), the dimensionless quadrupole moment ($\bar{Q} = Q/(J^2/M)$) and the dimensionless tidal deformability ($\Lambda = \lambda/M^5$) calculated using the numerical library LORENE and unified EoSs as described in Chap. 3. Earlier calculations of approximate universal relations for same observables of neutron stars were performed using polytropic EoSs [47, 48]. The $\bar{I}-\bar{Q}$ relationship is displayed

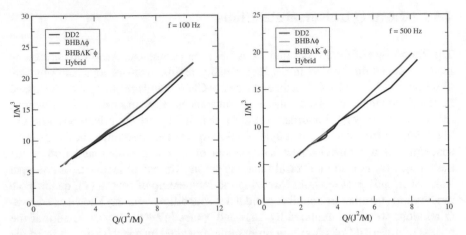

Fig. 4.3 The dimensionless moment of inertia is shown as a function of dimensionless quadrupole moment for different EoSs and rotational frequencies of $f = 100\,\mathrm{Hz}$ (left panel) and $f = 500\,\mathrm{Hz}$ (right panel). This is taken from Ref. [43] and reproduced with kind permission of the European Physical Journal

in the left panel of Fig. 4.3 for neutron stars rotating at a rotational frequency $f = 100\,\mathrm{Hz}$. It is found that this relationship is insensitive to nuclear (DD2), hyperon (BHB$\Lambda\phi$), and antikaon condensate (BHB$\Lambda K^-\phi$) EoSs [43]. However, this is not the same for the hybrid EoS. In this case, the hybrid EoS is undergoing a first order phase transition from the hadronic matter described by the DDRH model to the quark matter based on the nonlocal NJL model [57]. The hybrid EoS in Fig. 4.3 shows a clear deviation from the universality. This proves that a first order hadron-quark phase transition weakens the universality of $\bar{I}$–$\bar{Q}$ relation [43]. These results were shown for slowly rotating neutron stars with rotational frequency $f = 100\,\mathrm{Hz}$ [43]. This calculation was also repeated at higher rotational frequencies [43]. We present the $\bar{I}$–$\bar{Q}$ relation for a higher rotational frequency $f = 500\,\mathrm{Hz}$ in the right panel of Fig. 4.3. The outcome, in this case, is the same as that of $f = 100\,\mathrm{Hz}$.

Compositions of neutron stars might be modified due to the change in the rotation rate [58]. Particularly this could impact the hadron-quark phase transition in rotating neutron stars. As a neutron star rotates fast, the matter at the center would be diluted. A hadron-quark phase transition might not occur if the central density does not exceed the threshold value for the phase transition. This was indeed observed for the hadron-quark phase transition where the mixed phase was narrow in extent and the approximate universality was retained in fast rotating neutron stars [43]. The $\bar{I}$–$\bar{Q}$ relation is studied with the hybrid EoS at a higher frequency $f = 500\,\mathrm{Hz}$ to probe this effect. The deviation from the approximate universality still exists at higher rotational frequencies for the hybrid EoS. This may be explained by the fact that the hybrid EoS has a much wider hadron-quark mixed phase as evident from Fig. 3.6 in Chap. 3.

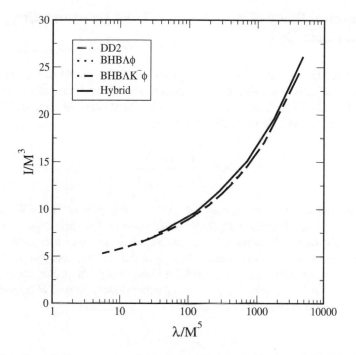

Fig. 4.4 The dimensionless moment of inertia is shown as a function of dimensionless tidal deformability for different EoSs. This is taken from Ref.[43] and reproduced with kind permission of the European Physical Journal

Figure 4.4 depicts the relation of the dimensionless moment of inertia ($\bar{I}$) with the dimensionless tidal deformability Λ for the same set of EoSs. An approximate universal relation in $\bar{I}$-Λ is obtained for all hadronic EoSs. Like Fig. 4.3, the results of the hybrid EoS deviate from the approximate universal relation. Figures 4.3 and 4.4 together result in I-Love-Q approximate universal relations for hadronic EoSs but this is no longer valid for EoSs with the wider mixed phase in first order phase transitions. Approximate universal relations among I and Q were also studied in rotating protoneutron stars [59, 60].

The approximate universality is a riddle! All EoSs in the sub-saturation density behave in the same fashion because the neutron star matter below the saturation density is well constrained by nuclear physics experiments. This might lead to the approximate universal relations in that density regime. However, many new degrees of freedom might appear in the form of hyperons, the antikaon condensate, and quarks at high densities. As a result, there would be appreciable deviations among EoSs including those exotic forms of matter as evident from our selections of EoS in this section. There is no credible explanation as to what is responsible for the approximate universality in the high density regime. There was some attempt to explain this approximate universality invoking the isodensity contours in neutron stars which are approximately elliptically self-similar [48].

4.5 Inspiral Phase of BNS Merger, Tidal Deformability, and Cold EoS

An inspiraling binary can be treated using two Newtonian point masses to the lowest order. Gravitational wave signals from inspiraling neutron stars in binaries might contain the valuable information about the chirp mass and the tidal deformability that shed light on the cold EoS. The chirp mass is

$$
M = \frac{(m_1 m_2)^{\frac{3}{5}}}{(m_1 + m_2)^{\frac{1}{5}}} \,,
\tag{4.14}
$$

where m_1 and m_2 are the masses of binary components. The GW frequency increases as binary components come closer due to the shrinkage of the orbit. Consequently the tidally induced mass quadrupole moment of a neutron star starts becoming important. We have already seen how the relation between the tidal field and quadrupole moment defines the tidal deformability. The influence of the tidal effects on the GW phase grows significantly with the increasing GW frequency. The Fourier transform of the GW strain ($\tilde{h}$) is written as [38, 61]

$$
\tilde{h} = A f_{GW}^{-\frac{7}{6}} e^{i(\Psi_p + \Psi_t)} \,,
\tag{4.15}
$$

where A is the amplitude and Ψ_p and Ψ_t denote phases of the GW signal due to the point mass and tidal effects, respectively. Further the tidal correction to the GW phase is given by [38, 39]

$$
\Psi_t = \frac{3}{128(\pi M f_{GW})^{\frac{5}{3}}} \left[-\frac{39}{2} \tilde{\Lambda} (\pi M_T f_{GW})^{\frac{10}{3}} \right] \,,
\tag{4.16}
$$

where M_T is the total mass of the binary and the combined tidal deformability is defined by [38]

$$
\tilde{\Lambda} = \frac{16}{13 M_T^5} \left[\left(1 + \frac{12 m_2}{m_1} \right) \lambda_1 + \left(1 + \frac{12 m_1}{m_2} \right) \lambda_2 \right] \,,
\tag{4.17}
$$

where $M_T = m_1 + m_2$ and λ_1 and λ_2 are tidal deformabilities of binary components. The tidal correction to the GW phase in Eq. (4.16) is of the fifth post-Newtonian (5PN) order [39]. The combined tidal deformability further simplifies to the following form using Eq. (4.13),

$$
\tilde{\Lambda} = \frac{16}{13} \frac{(m_1 + 12 m_2) m_1^4 \Lambda_1 + (m_2 + 12 m_1) m_2^4 \Lambda_2}{(m_1 + m_2)^5} \,,
\tag{4.18}
$$

where Λ_1 and Λ_2 are the dimensionless tidal deformabilities corresponding to component masses m_1 and m_2.

The parameters of the GW source were extracted by matching the data with the post-Newtonian waveform models and employing a Bayesian analysis [15, 24, 32, 38]. Given the GW data, the main motivation is to determine the posterior probability density function (PDF). According to Bayes' theorem, the PDF that provides the information about source parameters (θ) correctly describing the GW data can be written as

$$P(\theta|\text{data}) = C P(\theta) P(\text{data}|\theta) , \qquad (4.19)$$

where C is a normalization constant. Here $P(\theta)$ represents priors for source parameters, $P(\text{data}|\theta)$ is the likelihood of getting the GW data in case of a signal with parameters θ existing in the data. A waveform model is needed to map the parameters to the signal in this analysis [15, 24, 32]. We discuss the properties of GW170817 obtained using the PhenomPNRT waveform model [24].

We were very lucky that it was possible to extract the chirp mass and the combined tidal deformability $\tilde{\Lambda}$ from the GW data of the first BNS merger GW170817 [15]. The chirp mass in the detector frame was obtained from the gravitational wave phase and the source chirp mass was estimated to be 1.186 ± 0.001 M$_\odot$ using the knowledge of the red-shift $z = 0.0099 \pm 0.0009$ of the host galaxy NGC 4993 [15, 62].

Unlike the chirp mass, the binary component masses are less constrained because of the degeneracy between the mass ratio $q = m_2/m_1$ and dimensionless aligned spin components along the orbital angular momentum χ_1 and χ_2. Two separate priors for the magnitudes of dimensionless spins were implemented in the analysis. The high spin implied to the prior $\chi \leq 0.89$ and the low spin prior represented $\chi \leq 0.05$ [15, 24]. The low spin case is consistent with the binary system like the double pulsar binary PSR J0737-3039. For the high spin case, the range of values of the mass ratio $q = 0.53$–1.0 and the values of m_1 and m_2 were found in the PhenomPNRT waveform model to be 1.36–1.89 M$_\odot$ and 1.00–1.36 M$_\odot$, respectively, whereas those values of component masses were $m_1 = 1.36$–1.60 M$_\odot$ and $m_2 = 1.16$–1.36 M$_\odot$ for the low spin case with $q = 0.73$–1.0 [24]. This analysis led to the 90% upper limit on the combined tidal deformability $\tilde{\Lambda}$ as 720 [24].

The PDF of tidal deformabilities Λ_1 and Λ_2 corresponding to the high and low mass components of GW170817 is displayed in Fig. 4.5 for the low spin case [24]. Furthermore, 90% (solid lines) and 50% (dashed lines) contours on the posterior distribution for four different waveform models are also shown in the plot. The waveform models differ from each other with respect to the treatment of the inspiral in absence of tidal effects, the inclusion of tidal effects, and spin induced quadrupole effects. The waveform models except the TaylorF2 model generated similar 90% upper bounds because the same tidal effects were incorporated in those waveform models [24]. The tidal deformabilities were allowed to vary independently in this analysis implying the use of different EoSs for the binary components [15, 24]. Predictions of the tidal deformability involving several representative piecewise

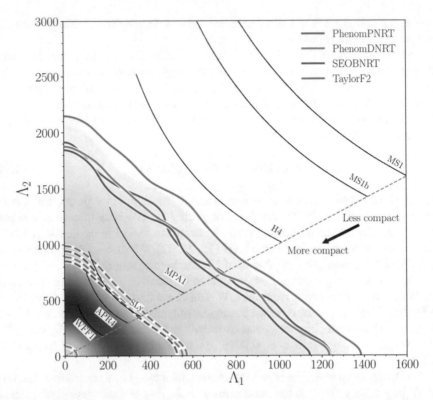

Fig. 4.5 The PDF for tidal deformabilities Λ_1 and Λ_2 is plotted along with 90% (solid lines) and 50% (dashed lines) confidence intervals for different waveform models and the low spin case. Black lines represent tidal deformabilities of different EoS models for which masses are computed using the PhenomPNRT waveform model, and terminate on $\Lambda_1 = \Lambda_2$ boundary. The blue shaded region indicates the PDF of precessing waveform model PhenomPNRT. Reprinted figure from Abbott [24] under the terms of the Creative Commons Attribution 4.0 International license. https://doi.org/10.1103/PhysRevX.9.011001

polytropic EoSs were shown in the discovery paper [15, 63]. All EoSs used here satisfy $\sim 2M_\odot$ neutron stars. However, several EoSs employed in the discovery paper of GW170817 are not compatible with up-to-date nuclear physics inputs [30]. Neutron stars become more compact as one approaches from the top right corner toward the origin of the figure. It is noted that the constraints on the tidal deformabilities in Fig. 4.5 rule out EoSs resulting in less compact stars.

The dependence of $\tilde{\Lambda}$ on the mass ratio q for the fixed chirp mass is found by writing Eq. (4.18) in terms of q,

$$\tilde{\Lambda} = \frac{16}{13} \frac{(1 + 12q)\Lambda_1 + (q + 12)q^4\Lambda_2}{(1+q)^5} \ . \tag{4.20}$$

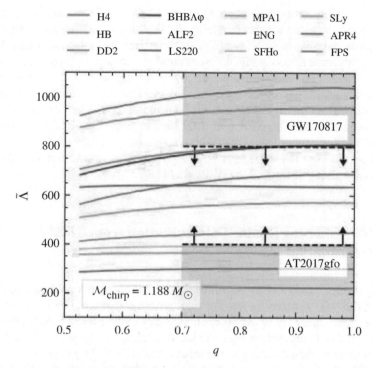

Fig. 4.6 The combined tidal deformability is plotted with the mass ratio q for different EoSs. This is taken from Ref.[64] and reproduced by permission of the American Astronomical Society (AAS). https://doi.org/10.3847/2041-8213/aaa402

Figure 4.6 shows the plot of $\tilde{\Lambda}$ with q for a set of EoSs. The color lines denoted different EoSs whereas the dashed lines imply the upper and lower bounds on $\tilde{\Lambda}$ as derived from the GW data of GW170817 and the EM counterpart AT 2017gfo. Too stiff or soft EoSs are beyond those limits and ruled out. On the other hand, the DD2 and BHB$\Lambda\phi$ EoSs lie just within the upper bound. It is observed that there is little variation of $\tilde{\Lambda}$ with q [31, 64, 65].

Soumi De et al. independently analyzed the GW data of GW170817 with the assumption that both neutron stars in the binary were described by the same EoS [33]. This indicates that the tidal deformabilities of binary components and the mass ratio may satisfy the following relation [33]

$$\frac{\Lambda_1}{\Lambda_2} \simeq q^6 . \tag{4.21}$$

This approximate correlation was also investigated by others [31, 66]. It was found that $q^6\Lambda_2/\Lambda_1$ deviated appreciably from the value of unity for $q \le 0.9$ [31]. The analysis of De et al. showed that the 90% upper bound on $\tilde{\Lambda}$ was reduced by 19% than that of the first LVC analysis [15, 24, 33]. Similar analysis assuming the same

EoS for both neutron stars was carried by the LVC [32]. In this analysis, the LVC exploited the universal relation involving the tidal deformability and compactness of a neutron star and the direct sampling parametrized EoSs. Consequently the area of 90% confidence region in Fig. 4.5 was found to shrink almost by a factor of 3.

GW170817 was followed by the discovery of the second BNS merger event GW190425 [18]. The chirp mass in GW190425 was estimated to be 1.44 ± 0.02 $M_\odot$. The component masses m_1 and m_2 at 90% credible intervals were in the range 1.60–1.87 $M_\odot$ and 1.46–1.69 $M_\odot$, respectively, for the low spin case $\chi < 0.05$ whereas those values for the high spin case $\chi < 0.89$ were 1.61–2.52 $M_\odot$ and 1.12–1.68 $M_\odot$. The value of $\tilde{\Lambda} < 1100$ for the low spin case in GW190425 is not as constraining of the EoS as GW170817 [18].

4.6   Neutron Star Radius Determination from Tidal Deformability

The knowledge of tidal deformabilities from BNS mergers could be exploited to estimate the radii of neutron stars. The tidal deformabilities of GW170817 together with measured masses of binary components offer a new avenue to estimate radii of neutron stars as evident from Eq. (4.13). Different strategies were adopted for the determination of radii of binary components in GW170817 by various groups [32–34, 66, 67]. We discuss those approaches for the determination of radii of neutron stars in GW170817.

The LVC reanalyzed the GW data of GW170817 in the Bayesian approach assuming both compact objects described by the same EoS that led to a correlation between Λ_1 and Λ_2 [32]. The PhenomPNRT waveform model and the low spin prior were adopted in this context. In this reanalysis, the source parameters in Eq. (4.19) might be expressed as $\theta = (\theta_p, \theta_{EoS})$, where parameters θ_p describing the compact objects in the binary as point masses, and θ_{EoS} are EoS dependent parameters coming into play due to tidal deformabilities. Further two EoS insensitive relations were used to implement the same EoS for both neutron stars and determine their radii from tidal deformabilities [32]. One such EoS insensitive relation was the tidal deformability (Λ) versus the compactness parameter (C) [68, 69]. The other relation was the antisymmetric tidal deformability (Λ_a) as a function of mass ratio and the symmetric tidal deformability (Λ_s), i.e. $\Lambda_a(\Lambda_s, q)$ with $\Lambda_a = (\Lambda_2 - \Lambda_1)/2$ and $\Lambda_s = (\Lambda_2 + \Lambda_1)/2$ [32]. This relation resulted in pairs of tidal deformabilities Λ_1 and Λ_2 corresponding to a large number of EoSs when Λ_s sampled uniformly in the range (0, 5000) [70, 71]. The $\Lambda - C$ relation was exploited to obtain the posterior for the mass and radius of each binary component.

It is also possible to sample the EoS directly instead of sampling tidal deformabilities as mentioned in the preceding paragraph. In this case, parametrized EoSs were preferred to the composition dependent traditional EoSs as discussed in Chap. 3. The piecewise polytropic and spectral parametrizations are being widely used in the analysis of GW data in BNS mergers [32, 33].

An adiabatic index (Γ) determines a unique EoS or vice versa as evident from the definition of the adiabatic index (Γ),

$$\Gamma = \frac{\varepsilon + p}{p} \frac{dp}{d\varepsilon} , \qquad (4.22)$$

or one can write

$$\frac{d\varepsilon}{dp} = \frac{\varepsilon + p}{p\Gamma} . \qquad (4.23)$$

The LVC adopted the spectral parametrization that represented $\log\Gamma(p, \gamma_i)$ as polynomials of $\log p$ where the adiabatic index (Γ) is a function of pressure (p) and free parameters of the EoS, $\gamma_i = (\gamma_0, \gamma_1, \gamma_2, \gamma_3)$ [72, 73]. With this particular representation, the EoS is obtained by integrating Eq. (4.23) [72]. The high density EoS derived in the spectral representation was merged with the Skyrme EoS, SLy, in the sub-saturation density [32, 74]. Next the EoS parameters γ_i were sampled uniformly over prior ranges with the constraint on the value of Γ in the range 0.6 to 4.5 such that this parametrization included a large number of EoSs [32]. All those EoSs satisfied the causality, thermodynamic stability $d\varepsilon/dp > 0$ and maximum neutron star mass of 1.97 $M_\odot$ at least [32]. Finally tidal deformabilities Λ_1 and Λ_2 of two neutron stars were computed for each sample and the waveform template was determined.

The posterior for the pressure as a function of the rest mass density was calculated employing the spectral parametrization of the EoS [32]. The pressure prior (orange) and posterior (blue) are shown with the rest mass density in Fig. 4.7. The light and dark blue regions imply 90% and 50% credible posterior intervals, respectively, whereas the orange lines demonstrate the 90% credible prior interval. Further horizontal lines correspond to the 90% credible intervals of the central pressure for the lighter (dotted) and heavier (dashed) binary components. Light grey lines are the results of H4, APR4, and WFF1 EoS models [75–77] from the top to the bottom along with that of the BHB$\Lambda\phi$ EoS [78]. The pressure at twice and six times the saturation density (ρ_{nuc}) are $3.5^{+2.7}_{-1.7} \times 10^{34}$ and $9.0^{+7.9}_{-2.6} \times 10^{35}$ dyne/cm^2 at the 90% confidence level, respectively [32]. It is observed that the pressure posterior moves away from the 90% credible prior area implying the preference for softer EoSs than the prior [32]. Additional degrees of freedom such as hyperons, Bose–Einstein condensates or quarks at higher densities make the EoS softer. For example, the BHB$\Lambda\phi$ EoS becomes softer with the onset of Λ hyperons and bends further away from the 90% credible posterior interval above $\sim 4\rho_{\mathrm{nuc}}$. It may be noted that the posterior pressure above the horizontal lines could not be reliable [32].

Figure 4.8 exhibits the posteriors for masses and areal radii of binary components in GW170817 estimated using the EoS insensitive relation and parametrized EoS. The bottom (top) posterior corresponding to the lighter (heavier) neutron star is denoted by orange (blue) contours. Some examples of mass-radius relationships corresponding to SLy, H4, APR4, WFF1, and MPA1 [74–77, 79, 80] along with that

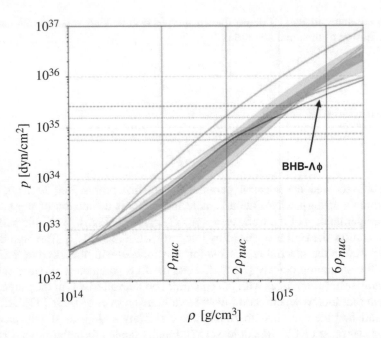

Fig. 4.7 The marginalized posterior and prior of the pressure (p) are plotted with mass density (ρ) along with 50% (90%) posterior credible interval in dark (light) blue and 90% prior credible interval in orange. Different EoSs are also overlaid on the plot. See the text for a detailed description of it. Reprinted figure with permission from Abbott et al. [32] ©2018 by the American Physical Society (APS). https://doi.org/10.1103/PhysRevLett.121.161101

of the BHB$\Lambda\phi$ EoS are also shown here. Two lines in the top left correspond to the Schwarzschild Black Hole (BH) and Buchdahl limits. Solid and dashed lines in one dimensional plots denote the posteriors and priors, respectively, whereas the vertical dotted lines provide the bounds of 90% credible intervals. The radii of neutron stars relevant for the range of component masses in GW170817 at the 90% level are the same, i.e. $R_1 = R_2 = 11.9 \pm 1.4$ km. It is evident from Fig. 4.8 that the radius ≤ 10.5 km in case of the soft WFF1 EoS or ≥ 14 km for the stiff H4 EoS are ruled out by this analysis. The other analysis using $\Lambda_a(\Lambda_s, q)$ and the EoS insensitive relation $\Lambda - C$ resulted in the radii $R_1 = 10.8^{+2.0}_{-1.7}$ km and $R_2 = 10.8^{+2.1}_{-1.5}$ km that could accommodate the soft WFF1 EoS [32].

De et al. reanalyzed the GW data of GW170817 in the Bayesian approach assuming both compact objects described by the same EoS that led to the correlation between Λ_1 and Λ_2 [33]. A large number of EoSs were generated based on the piecewise polytropic formulation [63]. Those EoSs satisfied experimental, observational, and theoretical constraints [33]. They constrained the radius $8.7 \leq R/\text{km} \leq 14.1$ in the 90% credible interval.

The radius ($R_{1.4}$) of a 1.4 $M_\odot$ neutron star was obtained as a function of the combined tidal deformability analytically in light of GW170817 [33, 66]. In

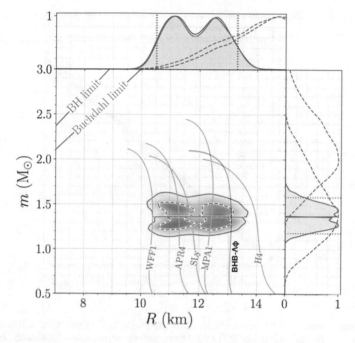

Fig. 4.8 Marginalized posteriors of masses and areal radii corresponding to binary components of GW170817 computed using the parametrized EoS and the EoS insensitive relation are shown here. The top (bottom) posterior in blue (orange) implies to the heavier (lighter) neutron stars. The solid and dashed lines in one dimensional plots represent the posteriors and priors, respectively. The bounds of 90% credible intervals are shown by the vertical dotted lines. Mass-radius relations of different EoSs are also overlaid on the plot along with that of the BHBΛϕ EoS. Reprinted figure with permission from Abbott et al. [32] ©2018 by the APS. https://doi.org/10.1103/PhysRevLett.121.161101

this approach, the radii corresponding to component masses of GW170817 were assumed to vary little, i.e. $R_1 \simeq R_2$ such that tidal deformabilities satisfied the correlation of Eq. (4.21). Finally they arrived at the expression of $R_{1.4}$ that is given by [66]

$$R_{1.4} \simeq (11.5 \pm 0.3 \text{ km}) \frac{M}{M_\odot} \left(\frac{\tilde{\Lambda}}{800} \right). \tag{4.24}$$

This semi-analytical approach was extended to find out the radii of neutron stars in the mass range $1.1 \, M_\odot \lesssim M \lesssim 1.6 \, M_\odot$ of binary components in GW170817 [67]. A large number of realistic EoSs recorded in Table 3.1 were used in this analysis. It was observed that the assumption $R_1 \simeq R_2 \simeq R$ was reasonable in the range of component masses of GW170817 for these realistic EoSs. The tidal deformability defined in Eq. (4.13) scales with the compactness parameter as C^{-6} where the quadrupolar Love number $k_2 \propto C^{-1}$. A plot of Λ with C^{-6} for all

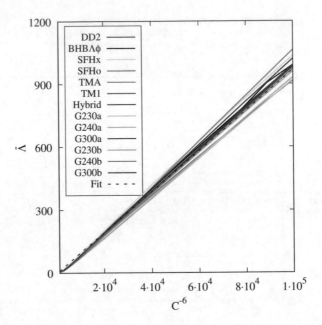

Fig. 4.9 The combined tidal deformability is plotted with C^{-6} where C is the compactness parameter. This is taken from Ref.[67] and reproduced by permission of the AAS. https://doi.org/10.3847/1538-4357/ab6a9e

those EoSs is shown in Fig. 4.9 and the proportionality constant of the linear fit is $a \simeq 0.00976$ [67]. After plugging in the linear fit, considering the assumption on radii and writing m_1 in terms of the chirp mass and mass ratio, the combined tidal deformability of Eq. (4.18) simplifies to [66, 67],

$$\tilde{\Lambda} = \frac{16a}{13} \times \left(\frac{R}{M}\right)^6 \times \frac{q^{8/5}}{(1+q)^{26/5}} \times \left[12 - 11q + 12q^2\right]. \qquad (4.25)$$

It has been already noted that $\tilde{\Lambda}$ weakly dependent on the mass ratio q as evident in Fig. 4.6. One can find that the dependence of the combined tidal deformability on the chirp mass resembles to that of the tidal deformability on mass in Eq. (4.13). We can write Eq. (4.25) as

$$\tilde{\Lambda} = a' \left(\frac{R}{M}\right)^6 . \qquad (4.26)$$

The values of a' corresponding to $q = 0.7$ and 1.0 are 0.0043 and 0.0042, respectively, and almost constant in the mass ratio range $q = 0.7$–1.0 of GW170817. Now the radii corresponding to component masses of GW170817 can be obtained from Eq. (4.25) as

$$R \simeq 3.669 \times \frac{M}{M_\odot} \times \tilde{\Lambda}^{1/6} \, \text{km} \, . \tag{4.27}$$

The chirp mass $M = 1.186$ and upper bound on $\tilde{\Lambda} = 720$ of GW170817 give the value of radius as 13.03 km.

There were several theoretical estimates of the radii of binary components following the determination of $\tilde{\Lambda}$ by different groups [31, 34, 65, 81]. The statistical analysis of E. R. Most et al. had some novel features in EoSs [34]. Firstly, one million parametrized EoSs were generated adopting the crusts models of Baym, Pethick, Sutherland and Negele and Vautherin up to density $0.08 \, \text{fm}^{-3}$, nucleon-nucleon chiral interactions in the density regime 0.08–$0.21 \, \text{fm}^{-3}$ and the perturbative Quantum Chromodynamics for asymptotically high densities as discussed in the supplemental material of [34]. In the intermediate densities, EoSs were described by piecewise polytropes and matched with two extreme densities. Besides the usual constraints on overall EoSs such as the causality, thermodynamic stability, compatibility with the lower bound of the neutron star maximum mass of $2.01 \pm 0.04 \, M_\odot$, this analysis imposed additional conditions on each EoS like the upper bound on $\tilde{\Lambda} < 800$ and the upper limit of $2.16 \, M_\odot$ on the maximum mass that was an outcome of the merger remnant of GW170817 collapsing into a black hole promptly [34]. They determined the radius of $1.4 \, M_\odot$ neutron star in the range $12.00 \leq R/\text{km} \leq 13.45$ at the 2σ confidence level. Most et al. also studied the phase transition in EoSs and obtained the stable branch beyond the neutron star branch and twin stars. In this case, the compact stars had smaller radii in the range $8.53 \leq R/\text{km} \leq 13.74$ at the 2σ confidence level.

It would be worth comparing these estimates of the radii of binary components in GW170817 with those of PSR J0030+0451 which were measured from the observations by the Neutron Star Interior Composition Explorer (NICER). Two different analyses of the NICER data on PSR J0030+0451 delivered the radius $12.71^{+1.14}_{-1.19}$ km corresponding to the mass $1.34^{+0.15}_{-0.16} \, M_\odot$ [82] and $13.02^{+1.24}_{-1.06}$ km with the mass $1.44^{+0.15}_{-0.14} \, M_\odot$ [83]. The mass of PSR J0030+0451 falls within the range of masses of binary components in GW170817. It is noted that the radii of neutron stars of GW170817 are compatible with those of PSR J0030+0451 except for smaller radii <11 km that were ruled out by the NICER measurement.

4.7   Hot and Neutrino-Trapped Merger Remnants and Finite Temperature EoSs

So far we have discussed the inspiral phase of BNS mergers and how the measurement of the tidal deformability in GW170817 could provide valuable information about the cold dense matter in neutron stars. The BNS mergers might give birth to rapidly rotating, hot and massive neutron star remnants that emit gravitational waves at higher frequencies ~ a few kilo Hertz. However, the LIGO and Virgo detectors did not record any postmerger GW signal in observed BNS mergers due to the lack of sensitivity of the detectors at such high frequencies [22]. The GW signals from neutron star remnants at such high frequencies would be rich in information of hot dense matter. There is now a proposal to build a high frequency GW interferometer named the Neutron Star Extreme Matter Observatory (NEMO) [84].

The precise outcome of a merger remnant very much depends on the EoS of dense matter and the maximum TOV mass. The merger remnants could undergo one of four possibilities such as (a) the prompt collapse to a black hole, (b) the formation of a HMNS that is supported by differential rotation and ultimately collapses to a BH in a second or so, (c) the birth of a supramassive neutron star (SMNS) that survives for ~10^4 s before collapsing to a BH and (d) a long-lived neutron star. In the absence of the postmerger kHz frequency GWs, it is not known what exactly happened to the compact remnant of GW170817. But the EM counterpart of GW170817 might provide some clues in this context.

4.7.1   Fate of BNS Merger Remnants

The ejected material generally originates from two sources in BNS mergers [27, 85]. The source of the dynamically ejected matter is the shock heating at the interface of two colliding neutron stars and tidal stripping. The other source of the ejected matter is the accretion disk around the compact object formed in the BNS merger. The ejecta due to the shock heating moves in the polar direction with a velocity 0.2–0.3 c and has a higher electron fraction ~0.3 and low opacity. The interaction of this material with neutrinos coming out of the HMNS makes it neutron-poor ejecta. Light r-process radioactive nuclei are synthesized in this Lanthanide-free ejecta. On the other hand, the neutron-rich matter ejected due to the tidal force in the equatorial plane has a low electron fraction <0.2 and high opacity due to the presence of heavy elements such as lanthanides synthesized in the r-process. The ejecta from the accretion disk contributes to the polar and equatorial components and has a wider electron fraction ~0.1–0.4.

The prediction of the kilonova (KN) was confirmed in the EM observations following the BNS merger in GW170817. The observations of the KN associated with GW170817 in Ultra-violet (UV)/Optical/Infrared(IR) might be explained in terms of two separate ejecta components. The blue KN was made of the neutron-poor matter that could describe the early UV and blue emissions. The red KN was responsible for the longer red/IR emission due to the decays of radioactive nuclei synthesized in the r-process in the neutron-rich material. It was estimated that the mass of the blue KN was $\sim 10^{-2}$ $M_\odot$ whereas the mass of the ejected neutron-rich material in the red KN was $\sim 5 \times 10^{-2}$ $M_\odot$. The total kinetic energy of the ejected material was $\sim 10^{51}$ ergs [23].

In case of a prompt collapse to a black hole, simulations of BNS mergers in numerical relativity produce a very negligible amount of ejected material that could not explain the blue KN mass as estimated in GW170817 [86]. Moreover, the black hole would launch a relativistic jet without any delay experiencing almost no material along its path in the prompt collapse scenario. These aspects in a prompt collapse are in tension with the observations. The observation of a sGRB was delayed by 1.7 s with respect to the time of the merger in GW170817. The delayed sGRB could be justified if the jet passes through the blue KN material, slows down and resulting in gamma ray emission. The estimated energy of the GRB was $\lesssim 10^{50}$ ergs [15].

The other scenario is the HMNS becoming an SMNS after losing its differential rotation completely. An SMNS might have huge rotational energy $\sim 10^{53}$ ergs. If it was a long-lived SMNS, it would have pumped a significant portion of this rotational energy into the merger surroundings [23, 86]. But this is not compatible with the energy budget of the GRB and KN associated with GW170817. It can be argued that the merger remnant of GW170817 was a short-lived HMNS that finally collapsed into a black hole.

4.7.2 Upper Bound on Maximum Mass of Neutron Stars from GW170817

The fate of the remnant depends on its total mass, the EoS, and the maximum TOV mass (M_{TOV}). The maximum mass of uniformly rotating neutron stars at the Kepler frequency is denoted by M_{Kep}. Neutron stars having masses greater than M_{TOV}, but less than M_{Kep}, are known as SMNSs. Any neutron star with mass above M_{Kep} is called a HMNS which is supported by the differential rotation.

We have already mentioned four possibilities about the fate of a BNS merger remnant. The remnant would promptly collapse to a black hole if the remnant mass (M_{rem}) is greater than a threshold mass (M_{thres}), i.e. $M_{rem} > M_{thres}$. The threshold mass is sensitive to the EoS and related to M_{TOV} by $M_{thres} = \eta M_{TOV}$. The value of η defines different phases of the remnant. Simulations of BNS mergers led to $\eta \sim 1.3$–1.6 in case of the prompt collapse to a black hole [23, 26, 87–91]. Long-lived SMNSs are born for $M_{rem} < M_{thres}$, where $\eta \sim 1.2$ was determined by investigating a large number of EoSs [49, 92, 93]. The condition $1.2 M_{TOV} \lesssim M_{rem} \lesssim 1.3$–$1.6 M_{TOV}$ might lead to the creation of a HMNS or a short-lived SMNS. The remnant would evolve to a stable neutron star for $M_{rem} \leq M_{TOV}$. The threshold mass was also found to depend on the radius of a non-rotating neutron as determined [91, 94],

$$M_{thres} = \left(-3.606 \frac{M_{TOV}}{R_{1.6}} + 2.38 \right) M_{TOV} , \qquad (4.28)$$

where $R_{1.6}$ is the radius of a 1.6 $M_\odot$ non-rotating neutron star.

The compact merger remnant in GW170817 was a short-lived HMNS that collapsed into a BH as evident from the EM observations. In this situation, it would be possible to place an upper limit on the maximum mass of the non-rotating neutron star. Several groups estimated the upper limit on M_{TOV} using the multimessenger observations and simulations in numerical relativity [23, 25–28]. It was assumed that the remnant was born as a HMNS, lost the support of differential rotation, and collapsed into a black hole close to the maximum mass of the uniformly rotating sequence at the Keplerian frequency. The total binary mass of GW170817 was $2.73^{+0.04}_{-0.01}$ $M_\odot$ for the low spin case. There were mass losses due to the ejected material and emissions of gravitational waves and neutrinos in the total binary mass [28]. The remnant mass was obtained by subtracting the mass loss from the total binary mass [25, 28]. The upper limit on M_{TOV} was obtained by equating the remnant mass with M_{thres} in the expression of a short-lived SMNS $M_{thres} \gtrsim 1.2 M_{TOV}$. Various calculations constrained the upper limit to $M_{TOV} \lesssim 2.17$ $M_\odot$ [23], $M_{TOV} \lesssim 2.16^{+0.17}_{-0.15}$ $M_\odot$ [25], $M_{TOV} \lesssim 2.28 \pm 0.23$ $M_\odot$ [26], and $M_{TOV} \lesssim 2.3$ $M_\odot$ [28]. Recently the LVC reported $M_{TOV} \leq 2.32$ $M_\odot$ based on the Bayesian analysis [95].

Now we discuss the implications of the upper limit of the maximum mass of neutron stars on the massive compact object of 2.52 $M_\odot$ in GW190425 [18] or the secondary component of $\sim 2.59^{+0.08}_{-0.09}$ $M_\odot$ in GW190814 [19]. All the results on the upper limit of the maximum mass of non-rotating neutron stars derived from the remnant in GW170817 are not compatible with massive compact object components in GW190425 and GW190814. In this context, we plot the mass versus central density of the non-rotating neutron stars (dash-dotted line) and the uniformly rotating neutron stars at the Keplerian limit (solid lines) for the DD2 EoS (left panel) and the BHB$\Lambda\phi$ EoS (right panel) in Fig. 4.10. The uniformly rotating models are calculated using the numerical library LORENE. These state-of-the-art EoSs even can not explain masses of non-rotating neutron stars $\gtrsim 2.42$ $M_\odot$. On the other hand, uniformly rotating stars might satisfy massive components in GW190425 or GW190814 for the DD2 EoS, but this does not happen even for the Keplerian sequence of the BHB$\Lambda\phi$ EoS. For the DD2 EoS, the secondary mass of GW190814 could be reached if it is rotating with a period of ~ 1.0 ms or smaller [96]. Other studies also noted that the mass of the secondary component of GW190814 was that of a sub-millisecond neutron star [97, 98]. In both panels of Fig. 4.10, we exhibit the sequences of uniformly rotating neutron stars at a fixed frequency of 1100 Hz that are represented by dotted lines and pass through the bounds of the secondary component mass in GW190814 as denoted by two horizontal lines in Fig. 4.10. It is evident that a stiffer EoS than the DD2 might result in the massive secondary component of GW190814 as a non-rotating neutron star.

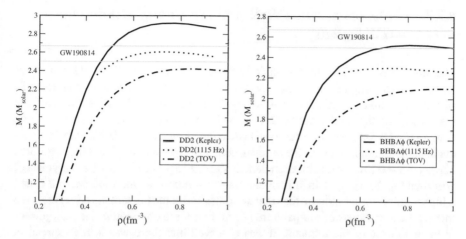

Fig. 4.10 Mass versus central density is shown for the sequences of non-rotating neutron stars (dash-dotted lines) and uniformly rotating neutron stars at the mass-shedding limit (solid lines) with DD2 EoS (left panel) and BHB$\Lambda\phi$ EoS (right panel) The sequences of uniformly rotating stars at a fixed frequency of 1100 Hz (dotted lines) are also exhibited here

However, there are some important issues that need to be addressed before we conclusively know the nature of the secondary component of GW190814—the heaviest neutron star or the lightest black hole. The central densities corresponding to the non-rotating sequence and uniformly rotating sequence at 1110 Hz for the DD2 EoS are ~5 times the saturation density or more. It would be difficult to consider the dense matter as nucleons-only matter at such high densities. It is expected that Λ hyperons might be populated first in this scenario. Consequently, the softer hyperon EoS is incompatible with the secondary component of GW190814 as evident from the right panel for the BHB$\Lambda\phi$ EoS. Secondly, another issue is the gravitational radiation driven Rossby (r)-mode instability in very fast rotating neutron stars with (sub)-millisecond periods due to the Chandrasekhar-Friedman-Schultz mechanism [99, 100]. The r-mode instability could be effectively damped by the hyperon bulk viscosity [101–106]. This implies that the r-mode might regulate the spin of neutron stars that rotate very fast at birth. No (sub)-millisecond pulsars have been detected so far and the fastest rotating pulsar has a spin frequency of 716 Hz [107]. The presence of hyperons for damping the r-mode in neutron stars rotating with (sub)-millisecond periods makes the mass of the secondary component of GW190814 incompatible with the hyperon EoS as evident from the BHB$\Lambda\phi$ EoS. The secondary component of GW190814 most likely might be a black hole.

4.7.3   Finite Temperature EoSs and Imprints of Exotic Matter in GW Signals

The hot, dense, and neutrino-trapped remnants formed in BNS mergers were extensively investigated in numerical relativity including finite temperature EoSs and the neutrino absorption and cooling by different groups [27, 108–114]. In those simulations of the total binary mass $\sim$2.6–3.0 $M_\odot$, the remnant was found to be composed of a massive compact object surrounded by a torous. It was reported that the maximum temperature of the remnant could reach $\sim$100 MeV, and the maximum density $\sim$5–10 times the saturation density [114, 115]. The entropy per baryon (s_B) was found to be $s_B \lesssim 1$ in Boltzmann unit in the core of the remnant and it had a higher value in the bulk of the remnant outside the unshocked core just after the merger. Higher values of temperature (T) and s_B were obtained in the later phases of the evolution of the remnant. It was observed that there was a large spread in temperature and entropy initially, but those became uniform at a later time.

The evolution of the remnant might be divided broadly into three time domains as gleaned from the simulations. Firstly it is driven by the emission of gravitational waves over 10–20 ms. Later this would be followed by the evolution under the viscosity on a time scale of $\sim$100 ms and neutrino cooling in 2–3 s [108, 110, 116–118]. The magneto-rotational instability (MRI) produces an effective viscosity in the remnant. Further, the magnetic winding during the differential rotation of the remnant leads to a competing effect with the MRI [117]. These effects together transport the angular momentum and remove the differential rotation that results in a rigidly rotating remnant [112].

Novel phases of matter made of hyperons, antikaon condensates and quarks might appear in the core of the remnant where the baryon density attains the maximum value according to numerical simulations. We highlight two such simulations involving hyperons and the hadron-quark phase transition and their imprints in gravitational waves [109, 113].

The BHBΛ EoS is being widely used in neutron star merger simulations after the discovery of gravitational waves from the BNS merger event GW170817 [109, 119–121]. Results of BNS merger simulations using the DD2 and BHB$\Lambda\phi$ EoS models and symmetric combinations of component masses 1.35 $M_\odot$ + 1.35 $M_\odot$, 1.4 $M_\odot$ + 1.4 $M_\odot$ and 1.5 $M_\odot$ + 1.5 $M_\odot$ are shown in Fig. 4.11 [109]. Here the gravitational wave strains and spectrograms assuming (+) polarization are plotted with time. Here $t = t_{mrg}$ denotes the time at the merger. The waveforms and the spectral content of the signals are shown in the top and bottom panels for the two EoS models, respectively. The Λ hyperon fraction remains low during the inspiral phase. However, the situation changes after the merger. In case of 1.5 $M_\odot$ + 1.5 $M_\odot$ neutron star merger with the BHB$\Lambda\phi$ EoS, the central density reaches a value as high as 4.5 n_0 at the merger [109]. Consequently, the population of Λ hyperons increases significantly. The waveforms of the DD2 and BHB$\Lambda\phi$ EoSs shown in the top panel of Fig. 4.11 deviate from each other after the merger. The waveform in the latter case is louder than that of the former. The distinctly different amplitude

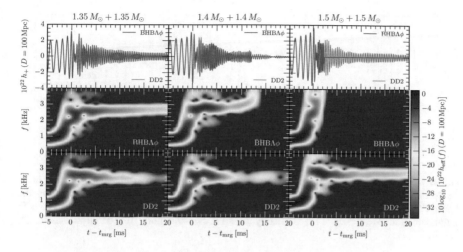

Fig. 4.11 Gravitational wave strains and spectrograms are plotted with time for BNS mergers with three combinations of component masses and the DD2 and BHB$\Lambda\phi$ EoS models [109]. This is reproduced by permission of the AAS. https://doi.org/10.3847/2041-8213/aa775f

modulation and phase evolution in the BHB$\Lambda\phi$ and DD2 waveforms make them distinguishable. It is evident from the bottom panels that the peak (f_2) frequencies for both EoS models exhibit little difference for the 1.35 $M_\odot$ + 1.35 $M_\odot$ neutron star merger. There is some difference in the case of 1.4 $M_\odot$ + 1.4 $M_\odot$ case with two EoSs. However, no f_2 frequency could be extracted from the 1.5 $M_\odot$ + 1.5 $M_\odot$ neutron star merger with the BHB$\Lambda\phi$ EoS as this system collapses promptly into a black hole [109]. This shows that the postmerger gravitational wave signals might carry the imprint of hyperons in BNS mergers.

The first order hadron-quark phase transition and its impact on the gravitational wave signal were also investigated in BNS merger simulations [113, 122–124]. E. R. Most et al. studied this problem in general relativistic BNS merger simulations including neutrino interactions and adopting an EoS undergoing the first order hadron-quark phase transition within the framework of the chiral mean field (CMF) model [113]. The same CMF model without quarks was employed to describe the hadronic EoS in the simulations. A large fraction of quarks was populated in the central region of the remnant at a later time of postmerger evolutions when the baryon density exceeded the threshold density of the phase transition. The appearance of quarks resulted in a softer EoS leading to the compression of the matter and a rise in the temperature of the central part of the remnant. It may be noted that the core of the remnant would be a relatively cold object ($T \lesssim 10\,\text{MeV}$) in absence of a phase transition. The results of these simulations, i.e. the GW strain, frequency, and phase difference are shown as a function of time in Fig. 4.12. Here $t = t_{mer}$ is the merger time. The left and right panels display the results of total binary masses 2.8 $M_\odot$ and 2.9 $M_\odot$ for the CMF EoSs with and without quarks, respectively [113]. The GW signals in the inspiral phase with and without quarks

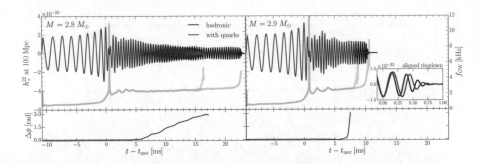

Fig. 4.12 Gravitational wave strain, frequency (top panel), and phase difference (bottom panel) are plotted with time for BNS mergers with the total mass of 2.8 $M_\odot$ and 2.9 $M_\odot$ and the hybrid EoS model and Chiral Mean Field hadronic EoS model of Ref.[113]. Reprinted figure with permission from Most et al. [113] ©2019 by the APS. https://doi.org/10.1103/PhysRevLett.122. 061101

do not show any difference because of the absence of quarks in this situation. The bottom plots in both panels exhibit the phase difference. In the case of the total binary mass 2.8 $M_\odot$, the phase difference starts rising after ∼5 ms due to the appearance of quarks in the remnant. This result is quite different from that of the simulations with the hyperon EoS where hyperons appear in the inspiral phase [109]. The phase difference is not noticeable for the simulations with the total binary mass 2.9 $M_\odot$ in the right panel of Fig. 4.12. The remnants were HMNS with and without quarks that finally collapsed into black holes. The remnants with quarks went to black holes earlier than the hadronic cases due to softening in the EoS in the former cases [113].

References

1. Einstein, A.: die feldgleichungen der gravitation, pp. 844–847. Sitzungsber. Preuss. Akad. Wiss., Berlin (1915)
2. Einstein, A.: Näherungsweise integration der feldgleichungen der gravitation, pp. 688–696. Sitzungsber. Preuss. Akad. Wiss., Berlin (1916)
3. Einstein, A., Rosen, N.: On gravitational waves. J. Franklin Inst. **223**(1), 43–54 (1937)
4. Hewish, A., Bell, S.J., Pilkington, J.D.H., Scott, P.F., Collins, R.A.: Observation of a rapidly pulsating radio source. Nature **217**, 709–713 (1968)
5. Hulse, R.A., Taylor, J.H.: Discovery of a pulsar in a binary system. Astrophys. J. Lett. **195**, L51–L53 (1975)
6. Symbalisty, E., Schramm, D.N.: Neutron star collisions and the r-process. Astrophys. Lett. **22**, 143–145 (1982)
7. Adhikari, R.X.: Gravitational radiation detection with laser interferometry. Rev. Mod. Phys. **86**(1), 121–151 (2014)
8. Nielsen, A.: Lecture notes on gravitational waves. J. Phys. Conf. Ser. **1263**, 012008 (2019)
9. Weber, J.: Evidence for discovery of gravitational radiation. Phys. Rev. Lett. **22**(24), 1320–1324 (1969)

10. Abramovici, A., et al.: LIGO – The laser interferometer gravitational-wave observatory. Science **256**, 325–333 (1992)
11. Bradaschia, C., et al.: Virgo: very wide band interferometric gravitational wave antenna. Nucl. Phys. B Proc. Suppl. **28**(1), 54–60 (1992)
12. Lück, H., et al.: The upgrade of GEO 600. J. Phys. Conf. Ser. **228**(1), 012012 (2010)
13. Joshi, B.C., et al.: Precision pulsar timing with the ORT and the GMRT and its applications in pulsar astrophysics. Astrophys. Astron. **39**(4), 51 (2018)
14. Abbott, B.P. et al.: Observation of gravitational waves from a binary black hole merger. Phys. Rev. Lett. **116**(6), 061102 (2016)
15. Abbott, B.P., et al.: GW170817: observation of gravitational waves from a binary neutron star inspiral. Phys. Rev. Lett. **119**(16), 161101 (2017)
16. Abbott, B.P., et al.: Multi-messenger observations of a binary neutron star merger. Astrophys. J. Lett. **848**(2), L12 (2017)
17. Abbott, B.P., et al.: Gravitational waves and gamma-Rays from a binary neutron star merger: GW170817 and GRB 170817A. Astrophys. J. Lett. **848**(2), L13 (2017)
18. Abbott, B.P., et al.: GW190425: observation of a compact binary coalescence with total mass. $\sim 3.4 M_{\odot}$ Astrophys. J. Lett. **892**(1), L3 (2020)
19. Abbott, R., et al.: GW190814: gravitational waves from the coalescence of a 23 Solar Mass black hole with a 2.6 solar mass compact object. Astrophys. J. Lett. **896**(2), L44 (2021)
20. Abbott, B.P. et al.: Observation of gravitational waves from two neutron star-black hole coalescences. Astrophys. J. Lett. **915**(1), L5 (2021)
21. Metzger, B.D., et al.: Electromagnetic counterparts of compact object mergers powered by the radioactive decay of r-process nuclei. Mon. Not. R. Soc. **406**, 2650–2662 (2010)
22. Abbott, B.P., et al.: Search for gravitational waves from a long-lived remnant of the binary neutron star merger GW170817. Astrophys. J. **875**, 160 (2019)
23. Margalit, B., Metzger, B.D.: Constraining the maximum mass of neutron stars from multi-messenger observations of GW170817. Astrophys. J. Lett. **850**, L19 (2017)
24. Abbott, B.P.: Properties of the binary neutron star merger GW170817. Phys. Rev. X **9**, 011001 (2019)
25. Rezzolla, L., Most, E.R., Weih, L.R.: Using gravitational-wave observations and quasi-universal relations to constrain the maximum mass of neutron stars. Astrophys. J. Lett. **852**, L25 (2018)
26. Ruiz, M., Shapiro, S.L., Tsokaros, A.: GW170817, general relativistic magneto-hydrodynamic simulations, and the neutron star maximum mass. Phys. Rev. D **97**, 021501 (2018)
27. Shibata, M., et al.: Modeling GW170817 based on numerical relativity and its implications. Phys. Rev. D **96**, 123012 (2017)
28. Shibata, M., Zhou, E., Kiuchi, K., Fujibayashi, S.: Constraint on the maximum mass of neutron stars using GW170817 event. Phys. Rev. D **100**, 023015 (2019)
29. Fonseca, E., et al.: Refined mass and geometric measurements of the high-mass PSR J0740+6620. Astrophys. J. Lett. **915**, L12 (2021)
30. Banik, S., Bandyopadhyay, D.: Dense matter in neutron star: Lessons from GW170817. In: Kirsch, J., Schramm, S., Steinheimer-Froschauer, J., Stöcker, H. (eds.) Discoveries at the Frontiers of Science, FIAS Interdisciplinary Science Series, pp. 85–94. Springer, Heidelberg (2020)
31. Bhat, S.A., Bandyopadhyay, D.: Neutron star equation of state and GW170817. J. Phys. G **46**, 014003 (2019)
32. Abbott, B.P., Abbott, R., Abbott, T.D., et al.: GW170817: Measurements of neutron star radii and equation of state. Phys. Rev. Lett. **121**, 161101 (2018)
33. De, S., Finstad, D., Lattimer, J.M., et al.: Tidal deformabilities and radii of neutron stars from the observation of GW170817. Phys. Rev. Lett. **121**, 091102 (2018); Erratum: Phys. Rev. Lett. **121**, 259902 (E) (2018)
34. Most, E.R., Weih, L., Rezzolla, L., Schaffner-Bielich, J.: New constraints on radii and tidal deformabilities of neutron stars from GW170817. Phys. Rev. Lett. **120**(26), 261103 (2018)

35. Thorne, K.S.: Multipole expansions of gravitational radiation. Rev. Mod. Phys. **52**, 299–340 (1980)
36. Thorne, K.S.: Tidal stabilization of rigidly rotating, fully relativistic neutron stars. Phys. Rev. D **58**, 124031 (1998)
37. Hinderer, T.: Tidal Love numbers of neutron stars. Atrophys. J. **677**, 1216 (2008); Erratum: Astrophys. J. **697**, 964 (2009)
38. Chaves, A.G., Hinderer, T.: Probing the equation of state of neutron star matter with gravitational waves from binary inspirals in light of GW170817: a brief review. J. Phys. G **46**, 123002 (2019)
39. Flanagan, E.E., Hinderer, T.: Constraining neutron-star tidal Love numbers with gravitational-wave detectors. Phys. Rev. D **77**, 021502 (2008)
40. Hinderer, T., Lackey, B.D., Lang, R.N., Read, J.S.: Tidal deformability of neutron stars with realistic equations of state and their gravitational wave signatures in binary inspiral. Phys. Rev. D **81**, 123016 (2010)
41. Regge, T., Wheeler, J.A.: Stability of a Schwarzschild singularity. Phys. Rev. **108**, 1063 (1957)
42. Thorne, K.S., Campolatto, A.: Non-radial pulsation of general-relativistic stellar models. I. Analytic analysis for $\ell \geq 2$. Astrophys. J. **149**, 591 (1967)
43. Bandyopadhyay, D., Bhat, S., Char, P., Chatterjee, D.: Moment of inertia, quadrupole moment, Love number of neutron star and their relations with strange-matter equations of state. Eur. Phys. J. A **54**, 26 (2018)
44. Damour, T., Nagar, A.: Relativistic tidal properties of neutron stars. Phys. Rev. D **80**, 084035 (2009)
45. Char, P., Banik, S.: Massive neutron stars with antikaon condensates in a density-dependent hadron field theory. Phys. Rev. C **90**, 015801 (2014)
46. Malik, T., Banik, S., Bandyopadhyay, D.: Equation-of-state table with hyperon and antikaon for supernova and neutron star merger. Astrophys. J. **910**, 96 (2021)
47. Yagi, K., Yunes, N.: I-Love-Q: unexpected universal relations for neutron stars and quark Stars. Science **341**, 365–368 (2013)
48. Yagi, K., Yunes, N.: Approximate universal relations for neutron stars and quark stars. Phys. Rep. **681**, 1–72 (2017)
49. Breu, C., Rezzolla, L.: Maximum mass, moment of inertia and compactness of relativistic stars. Mon. Not. R. Astron. Soc. **459**(1), 646–656 (2016)
50. Lenka, S.S., Char, P., Banik, S.: Critical mass, moment of inertia and universal relations of rapidly rotating neutron stars with exotic matter. Int. J. Mod. Phys. **26**, 1750127 (2017)
51. Bozzola, G., Stergioulas, N., Bauswein, A.: Universal relations for differentially rotating relativistic stars at the threshold to collapse. Mon. Not. R. Astron. Soc. **471**(3), 3557–3564 (2018)
52. Carson, Z., Chaziionnou, K., Yagi, K., Yunes, N: Equation-of-state insensitive relations after GW170817. Phys. Rev. D **99**(8), 083016 (2019)
53. Doneva, D.D., Yazadjiev, S.S., Stergioulas, N., Kokkotas, K.D.: Breakdown of I-Love-Q universality in radidly rotating relativistic stars. Astrophys. J. **781**(1), L6 (2014)
54. Haskell, B., Ciolfi, R., Pannarale, F., Rezzolla, L.: On the universality of I-Love-Q relations in magnetized neutron stars. Mon. Not. R. Astron. Soc. Lett. **438**(1), L71–L75 (2014)
55. Chakrabarti, S., Delsate, T., Gürlebeck, N., Steinhoff, J.: I − Q relation for rapidly rotating neutron Stars. Phys. Rev. Lett. **112**(20), 201102 (2014)
56. Lattimer, J.M., Schutz, B.F.: Constraining the equation of state with moment of inertia measurements. Astrophys. J. **629**, 979–984 (2005)
57. Orsaria, M., Rodrigues, H., Weber, F., Contrera, G.A.: Quark deconfinement in high-mass neutron stars. Phys. Rev. D **89**(1), 015806 (2014)
58. Mellinger, R.D. Jr, Weber, F., Spinella, W., Contrera, G.A., Orsaria, M.G.: Quark deconfinement in rotating neutron stars. Universe **3**(1), 5 (2017)
59. Martinon, G., Maselli, A., Gualtieri, L., Ferrari, V.: Rotating protoneutron stars: Spin evolution, maximum mass, and I-Love-Q relations. Phys. Rev. D **90**(6), 064026 (2014)

60. Marques, M., Oertel, M., Hempel, M., Novak, J.: New temperature dependent hyperonic equation of state: Application to rotating neutron star models and I–Q relations. Phys. Rev. C **96**, 045806 (2017)
61. Cutler, C., Flanagan, E.E.: Gravitational waves from merging compact binaries: how accurately can one extract the binary's parameters from the inspiral waveform? Phys. Rev. D **49**(6), 2658–2697 (1994)
62. Abbott, B.P., Abbott, R., Abbott, T.D., et al.: The basic physics of the binary black hole merger GW150914. Ann. Phys. **529**(1–2), 1600209 (2017)
63. Read, J.S., Lackey, B.D., Owen, B.J., Friedman, J.L.: Constraints on a phenomenologically parametrized neutron-star equation of state. Phys. Rev. D **79**(12), 124032 (2009)
64. Radice, D., Perego, A., Zappa, F., Bernuzzi, S.: GW170817: joint constraint on the neutron star equation of state from multimessenger observations. Astrophys. J. Lett. **852**, L29 (2018)
65. Raithel, C., Özel, F., Psaltis, D.: Tidal deformability from GW170817 as a direct probe of the neutron star radius. Astrophys. J. Lett. **857**(2), L23 (2018)
66. Zhao, T., Lattimer, J.M.: Tidal deformabilities and neutron star mergers. Phys. Rev. D **98**(6), 063020 (2018)
67. Soma, S., Bandyopadhyay, D.: Properties of binary components and remnant in GW170817 using equations of state in finite temperature field theory models Astrophys. J. **890**(2), 139 (2020)
68. Maselli, A., Cardoso, V., Ferrari, V., Gualtieri, L., Pani, P.: Equation-of-state-independent relations in neutron stars. Phys. Rev. D **88**(2), 023007 (2013)
69. Urbanec, M., Miller, J.C., Stuchlík, Z.: Quadrupole moments of rotating neutron stars and strange stars. Mon. Not. R. Astron. Soc. **433**(3), 1903–1909 (2013)
70. Yagi, K., Yunes, N.: Binary love relations. Class. Quant. Grav. **33**(13), 13LT01 (2016)
71. Chatziioannou, K., Haster, C.-J., Zimmerman, A.: Measuring the neutron star tidal deformability with equation-of-state-independent relations and gravitational waves. Phys. Rev. D **97**(10), 104036 (2018)
72. Lindblom, L.: Causal representations of neutron-star equations of state. Phys. Rev. D **97**(12), 123019 (2018)
73. Lindblom, L.: Spectral representations of neutron-star equations of state. Phys. Rev. D **82**(10), 103011 (2010)
74. Douchin, F., Haensel, P.: A unified equation of state of dense matter and neutron star structure. Astron. Astrophys. **380**, 151–167 (2001)
75. Lackey, B., Nayyar, M., Owen, B.: Observational constraints on hyperons in neutron stars. Phys. Rev. D **73**(2), 024021 (2006)
76. Akmal, A., Pandharipande, V.R., Ravenhall, D.G.: Equation of state of nucleon matter and neutron star structure. Phys. Rev. C **58**(3), 1804–1828 (1998)
77. Wiringa, R.B., Fiks, V., Fabrocini, A.: Equation of state for dense nucleon matter. Phys. Rev. C **38**(2), 1010–1037 (1988)
78. Banik, S., Hempel, M., Bandyopadhyay, D.: New hyperon equations of state for supernovae and neutron stars in density-dependent hadron field theory. Astrophys. J. Suppl. **214**(2), 22 (2014)
79. Müther, A., Prakash, M., Ainsworth, T.L.: The nuclear symmetry energy in relativistic Brueckner-Hartree-Fock calculations. Phys. Lett. B **199**(4), 469–474 (1987)
80. Müller, H., Serot, B.D.: Relativistic mean-field theory and the high-density nuclear equation of state. Nucl. Phys. A **606**, 508–537 (1996)
81. Fattoyev, F.J., Piekarewicz, J., Horowitz, C.: Neutron skins and neutron stars in the multimessenger era. Phys. Rev. Lett. **120**(17), 172702 (2010)
82. Riley, T.E., et al.: A NICER view of PSR J0030+0451: millisecond pulsar parameter estimation. Astrophys. J. Lett. **887**(1), L21 (2019)
83. Miller, M.C., et al.: PSR J0030+0451 mass and radius from NICER data and implications for the properties of neutron star matter. Astrophys. J. Lett. **887**(1), L24 (2019)
84. Ackley, K., et al.: Neutron star extreme matter observatory: A kilohertz-band gravitational-wave detector in the global network. Pub. Astron. Soc. Aus. **37**, e047 (2020)

85. Baiotti, L., Rezzolla, L.: Binary neutron star mergers: a review of Einstein's richest laboratory. Rep. Prog. Phys. **80**(9), 096901 (2017)
86. Gill, R., Nathanail, A., Rezzolla, L.: When did the remnant of GW170817 collapse to a black hole? Astrophys. J. **876**(2), 139 (2019)
87. Shibata, M.: Constraining nuclear equations of state using gravitational waves from hyper-massive neutron stars. Phys. Rev. Lett. **94**(20), 201101 (2005)
88. Shibata, M., Taniguchi, K.: Merger of binary neutron stars to a black hole: disk mass, short gamma-ray bursts, and quasinormal mode ringing. Phys. Rev. D **73**(6), 064027 (2006)
89. Baiotti, L., Giacomazzo, B., Rezzolla, L.: Accurate evolutions of inspiralling neutron-star binaries: prompt and delayed collapse to a black hole. Phys. Rev. D **78**(8), 084033 (2008)
90. Hotokezaka, K., Kyutoku, K., Okawa, H., Shibata, M., Kiuchi, K.: Binary neutron star mergers: dependence on the nuclear equation of state. Phys. Rev. D **83**(12), 124008 (2011)
91. Bauswein, A., Baumgarte, T.W., Janka, H.T.: Prompt merger collapse and the maximum mass of neutron stars. Phys. Rev. Lett. **111**(11), 131101 (2013)
92. Cook, G.B., Shapiro, S.L., Teukolsky, S.A.: Rapidly rotating polytropes in general relativity. Astrophys. J. **422**, 227 (1994)
93. Cook, G.B., Shapiro, S.L., Teukolsky, S.A.: Rapidly rotating neutron stars in general relativity: realistic equations of state. Astrophys. J. **424**, 823 (1994)
94. Bauswein, A., Just, O., Janka, H.-T., Stergiolous, N.: Neutron-star radius constraints from GW170817 and future detections. Astrophys. J. **850**(2), L34 (2017)
95. Abbott, B.P., et al.: Model comparison from LIGO–Virgo data on GW170817's binary components and consequences for the merger remnant. Class. Quan. Grav. **37**(4), 045006 (2020)
96. Tsokaros, A., Ruiz, M., Shapiro, S.L.: GW190814: spin and equation of state of a neutron star companion. Astrophys. J. **905**(1), 48 (2020)
97. Most, E.R., et al.: A lower bound on the maximum mass if the secondary in GW190814 was once a rapidly spinning neutron star. Mon. Not. R. Astron. Soc. **499**(1), L82 (2020)
98. Biswas, B et al.: GW190814: on the properties of the secondary component of the binary. Mon. Not. R. Astron. Soc. **505**(2), 1600 (2021)
99. Chandrasekhar, S.: Solutions of two problems in the theory of gravitational radiation. Phys. Rev. Lett. **24**(11), 611–615 (1970)
100. Friedman, J.L., Schutz, B.F.: Secular instability of rotating Newtonian stars. Astrophys. J. **222**, 281–296 (1978)
101. Lindblom, L., Owen, B.J., Morsink, S.M.: Gravitational radiation instability in hot young neutron stars. Phys. Rev. Lett. **80**(22), 4843–4846 (1998)
102. Lindblom, L., Owen, B.J., Ushominsky, G.: Effect of a neutron-star crust on the r-mode instability. Phys. Rev. D **62**(8), 084030 (2000)
103. Andersson, N.: Gravitational waves from instabilities in relativistic stars. Clas. Quan. Grav. **20**(7), R105–R144 (2003)
104. Stergioulas, N.: Rotating stars in relativity. Liv. Rev. Rel. **6**(1), 3 (2003)
105. Nayyar, M., Owen, B.J.: R-modes of accreting hyperon stars as persistent sources of gravitational waves. Phys. Rev. D **73**(8), 084001 (2006)
106. Chatterjee, D., Bandyopadhyay, D.: Effect of hyperon-hyperon interaction on bulk viscosity and r-mode instability in neutron stars. Phys. Rev. D **74**(2), 023003 (2006)
107. Hessels, J.W.T., et al.: A radio pulsar spinning at 716 Hz. Science **311**, 1901 (2006)
108. Sekiguchi, Y., Kiuchi, K., Kyutoku, K., Shibata, M.: Gravitational waves and neutrino emission from the merger of binary neutron stars. Phys. Rev. Lett. **107**(5), 051102 (2011)
109. Radice, D., Bernuzzi, S., Del Pozzo, W., Roberts, L.K., Ott, C.D.: Probing extreme-density matter with gravitational-wave observations of binary neutron star merger remnants. Astrophys. J. Lett. **842**, L10 (2017)
110. Radice, D., Perego, A., Bernuzzi, S., Zhang, B.: Long-lived remnants from binary neutron star mergers. Mon. Not. R. Astron. Soc **481**(3), 3670–3682 (2018)
111. Breschi, M., et al.: Kilohertz gravitational waves from binary neutron star remnants: time-domain model and constraints on extreme matter. Phys. Rev. D **100**(10), 104029 (2019)

112. Cioffi, R., Kastaun, W., Kalinani, V.J., Giacomazzo, B.: First 100 ms of a long-lived magnetized neutron star formed in a binary neutron star merger. Phys. Rev. D **100**(2), 023005 (2019)

113. Most, E.R., et al.: Signatures of quark-hadron phase transitions in general-relativistic neutron-star mergers. Phys. Rev. Lett. **122**, 061101 (2019)

114. Endrizzi, A., et al.: Thermodynamics conditions of matter in the neutrino decoupling region during neutron star mergers. Eur. Phys. J. A **56**(1), 15 (2020)

115. Lalit, S., Mamun, M.A.A., Constantinou, C., Prakash, M.: Dense matter equation of state for neutron star mergers. Eur. Phys. J. A **55**(1), 10 (2019)

116. Kiuchi, K., Kyutoku, K., Sekiguchi, Y., Shibata, M.: Global simulations of strongly magnetized remnant massive neutron stars formed in binary neutron star mergers. Phys. Rev. D **97**(12), 124039 (2018)

117. Hotokezaka, K., et al.: Remnant massive neutron stars of binary neutron star mergers: Evolution process and gravitational waveform. Phys. Rev. D **88**(4), 044026 (2013)

118. Fujibayashi, S., Kiuchi, K., Nishimura, N., Sekiguchi, Y., Shibata, M.: Mass ejection from the remnant of a binary neutron star merger:viscous-radiation hydrodynamics study. Astrophys. J. **860**(1), 64 (2018)

119. Bauswein, A., Stergioulas, N., Janka, H. -T.: Revealing the high-density equation of state through binary neutron star mergers. Phys. Rev. D **90**(2), 023002 (2014)

120. Dietrich, T., et al.: CoRe database of binary neutron star merger waveforms. Clas. Quan. Grav. **35**(24), 24LT01 (2018)

121. Most, E.R., Pappenfort, L.J., Tottle, S.D., Rezzolla, L.: On accretion discs formed in MHD simulations of black hole neutron star mergers with accurate microphysics. Mon. Not. R. Astron. Soc. **506**(3), 3511–3526 (2021)

122. Bauswein, A., et al.: Identifying a first-order phase transition in neutron-star mergers through gravitational waves. Phys. Rev. Lett. **122**(6), 061102 (2019)

123. Weih, L.R., Hanauske, M., Rezzolla, L.: Postmerger gravitational-wave signatures of phase transitions in binary mergers. Phys. Rev. Lett. **124**(17), 171103 (2020)

124. Hanauske, M., Weih, L.R., Stöcker, H., Rezzolla, L.: Metastable hypermassive hybrid stars as neutron-star merger remnants: a case study. Eur. Phys. J. Sp. Top. **230**(2), 543–550 (2021)

Chapter 5
Synthesis of Heavy Elements in the Universe

Summary The answer to the question how elements heavier than iron-type nuclei got produced in the universe is presented in this chapter. The rates for the capture of neutrons by the heavy nuclei are compared to the beta decay rates of the unstable ones and then the slow, the rapid, and the p-processes are defined, followed by their detailed description. The theoretical and experimental inputs to the calculation of the abundances populated through the chain reactions for the r-process are discussed. The sites of the s-process as well as the r-process are listed and the recent knowledge of r-process production of elements during the merger of two neutron stars is presented highlighting the kilonova model and the model for studying the decompression of the ejected matter.

5.1 Different Modes of Nucleosynthesis: The s-, the r-, and the p-Processes

With 118 known elements and 80 of them having at least one stable isotope, the question that naturally arises is how were they produced in the universe. In the early universe after the big bang, the elements and isotopes involved were $^{1,2}H$, $^{3,4}He$, and ^{7}Li but not the ones heavier. During stellar burning after the formation of stars, hydrogen goes through fusion forming helium and releasing energy; then helium nuclei go through the triple alpha reactions producing carbon. This has been described in Sect. 2.3 of Chap. 2. But as noted earlier, these reactions stop with the formation of iron-type nuclei. This is because around $A=56$, the binding energy per nucleon reaches a maximum value and it decreases for higher values of A. Chapter 2 also introduced the concept of nuclear statistical equilibrium (NSE) and how at high temperatures the abundances of nuclei around $A=56$, reach equilibrium values where the rate of each strong or electromagnetic reaction equals the rate for

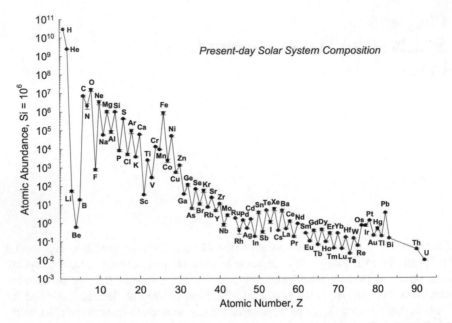

Fig. 5.1 The relative abundances of elements in our solar system. This is taken from Ref. [4] and reproduced with permission from the Springer

its inverse reaction. Thus during the advanced stellar evolution and the explosive end phases, elements from C up to the ones around Fe, i.e. Sc to Zn with $21 \leq Z \leq 30$ get created. As seen for nuclear statistical equilibrium, some small fractions of the stellar elements are heavier than the iron-type, and also during supernova evolution and shock formation in the matter deep inside these stars, the average A value increases beyond $A=56$; but these in no way can explain the observed abundances of the heavy elements. Thus the key question that remains unanswered is how do all the heavier elements up to Pb, Bi and others, are produced. This will be discussed in detail in this chapter highlighting how additional inputs from some very recent astrophysical observations are addressing the uncertainties and open questions in our understanding [1].

In Fig. 5.1 we show the relative abundances of elements in our solar system with H and He being the two most abundant ones. Here one can see the relatively smaller abundances of the elements from C, O to Fe, Co, Ni; but then the figure also shows the elements heavier, though with their values somewhat smaller, going right up to Pb, Bi and then Thorium and Uranium. Now let us discuss the physical processes that can produce them. As the nuclei start having more charge with a higher value of Z, the addition of charged particles to it, like protons or alphas, involves large Coulomb repulsions and as a result, such processes can take place only on comparatively lighter nuclei. But compared to that, the capture of neutrons can go ahead forming nuclei with one more neutron. Neutron capture cross sections, generally increase with decreasing energy. So in situations where neutron fluxes are not small and temperatures moderate, nucleosynthesis through neutron capture is a distinct possibility. Repeated capture of neutrons however will make the final nucleus radioactive and then through beta decay the neutron number will decrease by one with the atomic mass number A remaining unchanged. In such cases, the two competing quantities are the neutron capture rate and the beta decay rate. In 1957, Burbidge, Burbidge, Fowler and Hoyle (BBFH) [2] as well as Cameron [3] proposed a theory for the production of heavier elements through these processes and we describe that briefly here.

Let us consider the nucleus picking up a neutron through the reaction

$$(N, Z) + n \rightarrow (N + 1, Z) + \gamma . \tag{5.1}$$

If $(N+1, Z)$ is stable, it waits until it is able to absorb another neutron. Instead, if it is unstable, it may go through a beta decay. BBFH compared the neutron capture rates to the beta decay rates and proposed the rapid or r-process and the slow or s-process. The r-process is defined as the one with the neutron capture rate much faster than the rate of the beta decay. In that case, the nucleus $(N + 1, Z)$ picks up another neutron. On the other hand, the s-process has the n-capture rate much slower and then it decays by beta decay and as a result, the number of protons increases by one and the neutron number decrease by unity, i.e.

$$(N + 1, Z) \rightarrow (N, Z + 1) + e^{-} + \bar{\nu}_e . \tag{5.2}$$

Figure 5.2 gives a schematic picture of the nuclei getting heavier through multiple neutron capture and beta decays with the neutron number N along the x-axis and

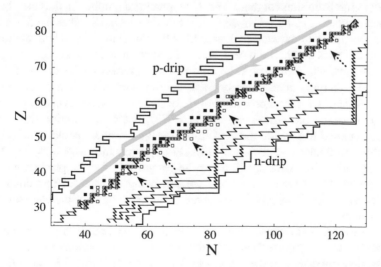

Fig. 5.2 A network of nuclei in the (N, Z) plane involved in the s-, r- and p-processes. For details see the text. Reprinted from [5] and ©2006 with permission from Elsevier

the proton number Z along the y-axis [5]. In the figure, a neutron capture takes a nucleus to the right and a beta decay takes the nucleus up making an angle of 135 degrees with the positive direction of the x-axis. The rightmost bold line shows the neutron drip line, i.e. the location of zero neutron separation energy line. The solid lines inside the n-drip are the r-process paths of a number of n-captures, followed by beta decay. The figure shows three such zigzag lines. The bold arrows indicate that in the absence of neutron sources the nuclei in the r-path go toward stability through multiple beta decays. The s- and r-isotopes are indicated by open squares whereas the p-isotopes are given by black squares. The elements start from the iron nuclei and go up to Bi. The thin black line close to the stable isotopes represents the up-streaming s-process path where once a nucleus picks up one neutron more than the stable one, moving one step horizontally (with fixed Z), it immediately beta decays back to stability. This is because, for the s-process, the beta decay rate is much faster than the n-capture rate as mentioned before. The upper leftmost bold line shows the proton drip line of zero proton separation energies and the capture of an extra proton vertically makes the nucleus unstable.

The important thing to note here is that most of the r-process nuclei are far from the stable and long-lived isotopes, i.e. away from nuclei for which a lot of nuclear properties and data were known experimentally. Consequently earlier not much nuclear information was available for the r-process nuclei. However in the last 15–20 years, experimental facilities are created or getting built, to study the specific nuclei involved in the r-process. Also through improved observation, their abundances in far-away stars are getting measured. Still compared to that, s-process nuclei are much better understood as their structural properties as well as the cross sections for the different reactions involving them, are known reasonably well.

The other issue that is crucial for a proper understanding of these processes is to identify the realistic sites for them. For, the s-process a number of sites are observed and the abundances of nuclei involved are calculated. On the other hand, for the r-process, one needs to have sites where a high flux of neutrons are available to make their capture rates rapid. The recent event of a neutron star-neutron star merger has shown the production of very large amounts of r-process nuclei in the site. Thus neutron star mergers or the merger of a neutron star with a black hole, provide possible scenarios for the r-process. For quite some time supernova explosions and the region outside the proto-neutron star (PNS) under the neutrino-driven wind, were conjectured to provide the conditions for the r-process nucleosynthesis. But presently one thinks that they are unlikely as major sites and it is felt that an r-process path with fission cycling can be responsible for the production of the heavy nuclei. These issues, however, are still being debated. In this chapter, we shall discuss in detail the physics involved in the production of the elements in the different sites.

For an overall understanding of the solar abundances of all the heavy nuclei, it is important to decompose them into three separate distributions involving the s-process, the r-process, and the p-process. This is shown in Fig. 5.3 where the three processes are individually shown. This separation depends somewhat on the models for calculating the synthesis of the isotopes in these processes. However one clearly

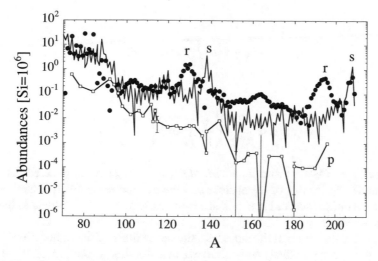

Fig. 5.3 The solar abundances of isotopes divided into s-, r- and p-processes. For details see the text. Reprinted from [5] and ©2006 with permission from Elsevier

observes specific peaks in these distributions as shown in the figure. There are peaks for the s-process at (i) $A= 88$ around the Sr isotopes (ii) $A = 138$ around the Ba isotopes and at (iii) $A= 206$–208 around the Pb isotopes. Similarly, for the r-process two clear peaks are observed at (i) $A= 128$–130 around the Xe isotopes and at (ii) $A =194$ around the Pt isotopes. Later when we discuss the models for describing the s- and the r-process, we shall explain why the abundances reach such peaks for these atomic mass numbers.

5.1.1   The s-Process

Experimentally it is not possible to get the neutron energy resolution comparable to the level separation of the heavy nuclei, excited to a few MeV. So the measurements give the cross section (σ) averaged over a range of energy which is sufficiently large to give σ as a smooth function of the energy. The actual dependence of σ on the velocity "v" for dominantly s-wave capture is $\sigma \propto v^{-1}$ at low energies. However for p-wave capture it is simply proportional linearly to "v". So assuming contribution from p-wave small, we can roughly take the averaged quantity, $< \sigma v >$ as constant. The neutron velocity "v" has a Maxwell–Boltzmann distribution involving the stellar temperature "T". It is often convenient to define an average cross section $< \sigma >$ and write [6, 7]

$$< \sigma >= \frac{< \sigma v >}{v_T},$$

(5.3)

with

$$< \sigma v >= \int_0^\infty \sigma v \phi(v) dv , \qquad (5.4)$$

where

$$\phi(v) dv = \frac{4}{\pi^{1/2}} \left(\frac{v}{v_T}\right)^2 exp\left[-(\frac{v}{v_T})^2\right] \frac{dv}{v_T} . \qquad (5.5)$$

Here $v_T = (\frac{2kT}{\mu_n})^{1/2}$ and $\mu_n = M_n M_A/(M_n + M_A)$ is the reduced mass with M_A and M_n the masses of the nucleus and the neutron respectively. Then one sees that for the nuclei involved $< \sigma >$ has a value close to $< \sigma_T >$, defined as the cross section at velocity v_T.

As per the definition of the s-process, the decay times of the radioactive isotopes involved are comparatively very short so that the time evolution of the number density $N_A(t)$ of the nucleus with the atomic number A can be written as [6–8]

$$\frac{dN_A}{dt} = - < \sigma v >_A N_n(t) N_A(t) + < \sigma v >_{A-1} N_n(t) N_{A-1}(t) , \qquad (5.6)$$

where $N_n(t)$ is the free neutron density, assumed uniform over the region considered. The first term in the RHS stands for the reduction in the rate for n- capture on the nucleus with nucleon number A and the second term is the addition due to the n-capture on the nucleus with nucleon number $A - 1$. The n-capture cross sections for the nuclei lighter than the iron group nuclei are much smaller than the ones for the iron group and beyond. Thus they can contribute only to much larger neutron flux and so it is reasonable to proceed with the iron group of elements as the seed nuclei for the synthesis of the heavier nuclei. Accordingly for the boundary condition for Eq. (5.6) one makes the simple but realistic assumption that at time $t = 0$, the nuclei are all $A = 56$ nuclei. These are the contents of a parametric model referred to as the canonical model of Clayton et al. [9] and later refined by Käppeler et al. [10]. This model also makes the simplifying assumption that the temperature and the neutron number density remain constant in time during the neutron irradiation process.

Thus the boundary condition is approximated as

$$N_A(0) = N_{56}(0) \quad for \quad A = 56 \quad , \qquad (5.7)$$

and

$$N_A(0) = 0 \quad for \quad A > 56 . \qquad (5.8)$$

Iliadis [7] writes the RHS of Eq. (1.7) as $f N_{56}^{Seed}(0)$ with a seed initial number $N_{56}^{Seed}(0)$ and a fraction f of them ($f \leq 1$) going through the process of synthesis. However, that does not change the procedure for the solution and we here take f

equal to one for simplicity. To study the evolution, one defines a new variable τ, named neutron exposure, as

$$d\tau = v_T N_n(t)dt , \tag{5.9}$$

so that $\tau = v_T \int N_A(t)dt$. The s-process chain so described ends at $A= 209$ as ^{209}Bi is the heaviest stable nucleus. The material at $A =210$ created with the nucleus ^{209}Bi capturing a neutron, however decays by α-particle emission to ^{206}Pb. This cycle terminates the s-process and the coupled set of differential equations can be written as [6] (writing σ_A instead of $< \sigma >_A$)

$$\frac{dN_{56}}{d\tau} = -\sigma_{56}N_{56} , \tag{5.10}$$

$$\frac{dN_A}{d\tau} = -\sigma_A N_A + \sigma_{A-1}N_{A-1} \quad for \quad 57 \le A \le 209 \quad A \ne 206 , \tag{5.11}$$

$$\frac{dN_{206}}{d\tau} = -\sigma_{206}N_{206} + \sigma_{205}N_{205} + \sigma_{209}N_{209} , \tag{5.12}$$

for the rest of the discussion let us limit ourselves to $56 \le A \le 205$ with the reminder that the termination solution is discussed in Clayton and Rassbach [11].

In reality, of course, the s-process occurs with a number of sources of free neutrons with different temperatures in different astrophysical conditions. But the simple assumptions of the canonical model lead us to analytical results which explain the basic features of the s-process.

Equation (5.11) shows for $\frac{dN_A}{d\tau}$ negative, one gets $N_A \ge (\frac{\sigma_{A-1}}{\sigma_A})N_{A-1}$ and for $\frac{dN_A}{d\tau}$ positive one has $N_A \le (\frac{\sigma_{A-1}}{\sigma_A})N_{A-1}$, or in other words, the solution of the equation is self-regulating and minimizes the difference between the two terms in the RHS. For nuclei away from shell closure, the cross sections have comparable values and one defines the "local approximation" for non-magic A values as

$$\sigma_A N_A = \sigma_{A-1}N_{A-1} . \tag{5.13}$$

This result is successful in estimating the abundances of the s-only (which have no r-process contribution) isotopes of many elements in agreement with their Solar System (SoS) abundances. The s-process yield of a nucleus is inversely proportional to the neutron capture cross section on it, as seen in Eq. (5.13). Though strictly not applicable near magic nuclei, still extending this feature, one expects s-process peaks for the stable nuclei ^{88}Sr, ^{138}Ba and ^{208}Pb with magic neutron numbers N=50, 82 and 126 respectively, as they have extremely small neutron capture cross sections. This indeed is observed for the SoS as described earlier.

Let us now discuss the analytical solutions of the Eqs. (5.10) and (5.11) for a specific case. We consider the situation where the fraction f of the total seed nuclei

N_{56}^{seed} are subjected to a neutron exposure which is exponential in the exposure τ, i.e.

$$p(\tau) = \frac{f N_{56}^{seed}}{\tau_0} exp(-\frac{\tau}{\tau_0}) .\tag{5.14}$$

The distribution above is normalized to $f N_{56}^{seed}$. The abundance for the nucleus with atomic number A as a result is

$$\overline{N_A(\tau_0)} = \frac{\int_0^\infty N_A(\tau)p(\tau)d\tau}{\int_0^\infty p(\tau)d\tau} = \int_0^\infty \frac{N_A(\tau)}{\tau_0} exp(-\frac{\tau}{\tau_0})d\tau .\tag{5.15}$$

Then the evolution Eqs. (5.10) and (5.11) for A=56 and A=57 become

$$\frac{dN_{56}(\tau)}{d\tau} = -N_{56}(\tau)\sigma_{56} ,\tag{5.16}$$

$$\frac{dN_{57}(\tau)}{d\tau} = -N_{57}(\tau)\sigma_{57} + N_{56}(\tau)\sigma_{56} .\tag{5.17}$$

As a result, for the exponential exposure cases, the analytical solutions are found to be [7]

$$\sigma_{56}\overline{N_{56}(\tau_0)} = \frac{f N_{56}^{seed}}{\tau_0} \frac{1}{[1 + \frac{1}{\tau_0\sigma_{56}}]} ,\tag{5.18}$$

$$\sigma_{57}\overline{N_{57}(\tau_0)} = \frac{f N_{56}^{seed}}{\tau_0} \frac{1}{[1 + \frac{1}{\tau_0\sigma_{56}}]} \frac{1}{[1 + \frac{1}{\tau_0\sigma_{57}}]} ,\tag{5.19}$$

and so on. The general solution is easily seen to be

$$\sigma_A\overline{N_A(\tau_0)} = \frac{f N_{56}^{seed}}{\tau_0} \Pi_{i=1}^A \frac{1}{[1 + \frac{1}{\tau_0\sigma_i}]} .\tag{5.20}$$

So with the knowledge of the capture cross section σ_A, a fit to the observed solar abundances gives us the values of the parameters f and τ_0. That helps one to identify the realistic conditions for the s-process.

The shortcomings of the canonical model are dealt with in the Multi-Event s-process (MES) [5, 12] model where improvements result from considering a number of neutron irradiations with different temperatures and neutron number densities.

For the s-process, when one considers the simple form of the evolution Eq. (5.6), it is assumed that all neutron capture rates are much slower than the β^- decay rates using the definition of s-process. However in some realistic cases, one comes across situations where a beta decay rate is comparable to the n-capture rates. At these points, the s-process path splits into two branches. These s-process branchings can be handled analytically for the simpler case when the neutron density $N_n(t)$ is constant with respect to time along with constant temperatures. For the actual expressions for the quantities like the ratio of the branching fractions to the two channels, we refer to Iliadis [7]. In other cases, like time varying neutron densities, one needs to use numerical integration methods.

Let us give the example of an s-process branching point caused by β^- decay from an excited isomer [8, 13]. For the mass number 176, the β^- decay from the r-process terminates at $^{176}_{70}Yb$ making $^{176}_{71}Lu$ and $^{176}_{72}Hf$ s-only nuclei (we remind the readers of the "Z" values for the three nuclei by giving them explicitly in the bottom left for each). Figure 5.4 depicts this case. Besides the long-lived ground state which has a half-life of 4.00×10^{10} y, ^{176}Lu has an isomeric state at an excitation energy of 0.123 MeV which has a half-life 3.664 h. Both states get populated by the $(n.\gamma)$ reaction on ^{175}Lu with known partial cross sections. The neutron capture on ^{175}Lu according to Beer and Käppeler[14] proceeds 36% of the time to ground state of ^{176}Lu and 64% to the isomeric state. For a neutron density of $N_n \sim 10^8$ neutrons/cm^3, the lifetime for the neutron capture by both the ground state and the isomeric state is of

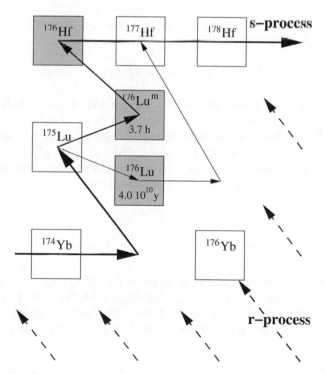

Fig. 5.4 The s-process neutron captures and beta decays in the Yb-Lu-Hf region (solid lines). For ^{176}Lu, the ground state and the isomer are shown separately. Reprinted figure with permission from [K. Langanke, G. Martinez-Pinedo, Rev. Mod. Phys. **75**, 819 (2003)] ©2003 by the American Physical Society (APS). https://doi.org/10.1103/RevModPhys.75.819

the order of 1 year [8]. Thus the ground state can be assumed to be comparatively stable to further n-capture while the isomeric state quickly decays to the stable isobar ^{176}Hf as shown in Fig. 5.4. Thus the s-process branching and the s-process matter flow, of ^{176}Lu is roughly 36% to ^{177}Lu and 64% to ^{176}Hf. The ^{177}Lu again β^- decays to ^{177}Hf. The ground state and the isomeric state of ^{176}Lu couple in the stellar photon bath through the excitation of an intermediate state at 0.838 MeV excitation, leading to a temperature dependent matter flow from the isomeric state to the ground state. The branching yields of the process can then be inverted to find the environment temperature. This was done by Doll et al. [15].

Finally, we briefly mention the present knowledge about the s-process sites. The s-process is made up of three separate components with distinct neutron exposures [16]: (a) the **weak** component with $A \leq 90$, from Fe to Sr, with mean neutron exposure $\tau \sim 0.06$ mbarn^{-1} with the site in the He and C burning shells of massive stars [17] (b) the **main** component with $90 \leq A \leq 204$ from Sr to Pb with $\tau \sim 0.3$ mbarn^{-1} taking place in low mass AGB stars (Asymptotic Giant Branch (AGB) stars found in the high-luminosity, low temperature region of the Hertzsprung-Russell (HR) diagram) and (c) the **strong** component which is for reproducing the solar Pb abundances, with $\tau \sim 7.0$ mbarn^{-1} expected in low mass, low metallicity AGB stars.

The r-process solar abundances $Y_A^\odot(r)$ are obtained from the s-process solar abundances $Y_A^\odot(s)$ by

$$Y_A^\odot(r) = Y_A^\odot - Y_A^\odot(s) , \tag{5.21}$$

where $Y_A^\odot$ is the SoS abundance for the nucleus with atomic number A. They have large uncertainties when $Y_A^\odot(s)$ is very close to $Y_A^\odot$ and then the model for the calculation of the s-process abundances should use the correct astrophysical environment over the galactic history prior to the formation of the solar system. Also in some cases for $A \leq 90$ small contributions from p-process and possibly 'α-process' [18, 19] in Eq. (5.21), need to be considered.

5.1.2 The r-Process

For the r-process, the resulting yield patterns are essentially controlled by the interplay among neutron capture, photo-disintegration, beta decay and possibly fission. For nuclei with proton number less than 80, fission is not important. Then the evolution of the number density for the nucleus (Z, A) can be expressed as

$$\frac{dN(Z, A)}{dt}$$
$$= N_n < v\sigma_{n,\gamma}(Z, A - 1) > N(Z, A - 1) + \lambda_{\gamma,n}(Z, A + 1)N(Z, A + 1)$$
$$+ \lambda_{\beta 0}(Z - 1, A)N(Z - 1, A) + \lambda_{\beta 1}(Z - 1, A + 1)N(Z - 1, A + 1)$$

$$+\lambda_{\beta 2}(Z - 1, A + 2)N(Z - 1, A + 2) + \lambda_{\beta 3}(Z - 1, A + 3)N(Z - 1, A + 3)$$

$$-N_n < v\sigma_{n,\gamma}(Z, A) > N(Z, A) \quad \lambda_{\gamma,n}(Z, A)N(Z, A)$$

$$-[\lambda_{\beta 0}(Z, A) + \lambda_{\beta 1}(Z, A) + \lambda_{\beta 2}(Z, A) + \lambda_{\beta 3}(Z, A)]N(Z, A), \tag{5.22}$$

where N_n is the neutron number density as earlier, $< v\sigma_{n,\gamma}(Z, A) >$ is the neutron capture rate averaged thermally, and $\lambda_{\gamma,n}(Z, A)$ is the photo-disintegration rate. The beta decay rates followed by the emission of 0,1,2,3 neutrons are given by $\lambda_{\beta 0}(Z - 1, A), \lambda_{\beta 1}(Z - 1, A + 1), \lambda_{\beta 2}(Z - 1, A + 2), \lambda_{\beta 3}(Z - 1, A + 3)$ respectively and the nuclei are referred to as (Z, A).

The above equation forms a reaction network. For r-process with both neutron capture and photo-disintegration taking place much faster than beta decay, the Eq. (5.22) becomes simpler and one sees that for $(n, \gamma) \rightleftharpoons (\gamma, n)$ equilibrium the ratio of the abundances of two neighboring isotopes is given by Qian [19] and Kar [20]

$$\frac{N(Z, A + 1)}{N(Z, A)} = N_n \left(\frac{2\pi\hbar^2}{m_u kT} \right)^{3/2} \left(\frac{A + 1}{A} \right)^{3/2} \frac{G(Z, A + 1)}{2G(Z, A)} exp \left[\frac{S_n(Z, A + 1)}{kT} \right].$$

$$\tag{5.23}$$

Here m_u is the atomic mass unit, $G(Z, A)$ is the nuclear partition function, and $S_n(Z, A)$ stands for the neutron separation energy of the nucleus (Z, A).

The conditions for nuclear statistical equilibrium (NSE) when all strong and electromagnetic reactions are balanced by their inverse reactions, are discussed in Chap. 2. Taking $N(Z, A + 1) \sim N(Z, A)$ and neglecting the small differences in the mass numbers and partition functions, one can get an expression for the neutron separation energy of the most abundant isotope [19]. It is seen that the most abundant isotope in different isotopic chains have approximately the same separation energy, S_n^0, which turns out to be 2–3 MeV for typical conditions during the r-process [19]. Due to the odd-even effect caused by nuclear pairing interaction, which makes the even neutron number nuclei more strongly bound, one often characterizes the most abundant isotope in the isotopic chain for a specific Z by a two neutron separation energy $S_{2n} = S_n(Z, A + 2) + S_n(Z, A + 1) \sim 2S_n^0$ [19].

It was seen that the $(n, \gamma) \rightleftharpoons (\gamma, n)$ equilibrium happens at $T \geq 2 \times 10^9$ K and $N_n \geq 10^{20}$/cc [21]. Later full network calculations got the equilibrium neutron density of 10^{20}/cm^3 for the temperature of 2×10^9 K and the equilibrium neutron density of 10^{28}/cm^3 for half that temperature [19]. The temperature for NSE is 6×10^9 K.

In this description fission cycling has not been taken into account. But if we consider the r-process for nuclei with $Z > 80$, then fission should be considered. In that case, the heaviest nucleus fissions and a cyclic flow occurs between this nucleus and its fission fragments in the presence of large neutron abundances. We discuss this more in the next section.

5.1.3 The p-Process

The abundances of the p-process nuclei for the SoS as shown in Fig. 2.3, are smaller than those of the s- and r-process nuclei on the average by a factor of $10^{-2} - 10^{-3}$. The idea of the process was proposed as early as 1957 by Burbidge et al. [2]. A detailed discussion of this is available in Arnould and Goriely [22]. Clearly, the n-deficient p-process nuclei cannot be produced in the neuron-capture chains of the s- and r-process. They can be reached through different routes like (a) repeated (p, γ) reactions making the seed nucleus proton-rich (b) repeated (γ, n) reactions reducing the neutron number with the number of protons remaining unchanged (c) (p, γ) reactions followed by a β^+ decay (d) (γ, n) followed by (γ, α) and then by a β^+ decay, etc. [23]. The relative importance of these various modes of reaching a p-nuclide depends sensitively on the temperature. For photo-disintegration rates with timescales comparable to the stellar evolution times, one needs temperature higher than 1.5×10^9 K. Also to avoid a very strong photoerosion, the temperature should be lower than 3.5×10^9 K and the hot phase for a short enough time. Along with these conditions for the proton capture reactions, one needs to have proton-rich layers for initiating the reactions. For understanding the physics of the process and for calculating the abundances, one needs to know the nuclear properties and the reaction cross sections of the nuclei involved lying between the stability line and the proton drip line. With the suggestions initially by Arnould, many studies of the p-process involve the explosion of the O-Ne-rich layers leading to the type II SN events. Another possible site is a SNIa explosion in a C-O white dwarf where once a mass close to the Chandrasekhar limit is reached through the accretion of matter in a binary system, a thermonuclear runaway of carbon develops leading to the explosion [22, 23]. For details of the production of the p-nuclides there, we refer to Arnould and Goriely [22].

5.2 Conditions for Production of Elements by the r-Process and the Sites

Before the discussions on the conditions for the production of the elements by the r-process and the sites, we briefly mention the waiting-point nuclei, the fission cycling and what is meant by freeze-out.

5.2.1 The Waiting-Point Nuclei

In Eq. (5.22) a steady flow happens when $\frac{dN}{dt} = 0$ for all the nuclei in the reaction network and the yields of these nuclei are given by a set of linear algebraic equations. The steady flow also satisfies

$$\lambda_\beta(Z-1)N(Z) = \lambda_\beta(Z)N(Z) , \tag{5.24}$$

where $N(Z) = \Sigma_A N(Z, A)$, $\lambda_\beta(Z) = \Sigma_A \lambda_\beta(Z, A)N(Z, A)/N(Z)$ and $\lambda_\beta(Z, A)$ is the total beta decay rate of the nucleus (Z,A). The expression for the beta decay rate is given in Chap. 2 in Eq. (2.14) with the nucleus in matter with temperature T. A special case of this steady flow r-process arises is when one has $(n, \gamma) \rightleftharpoons (\gamma, n)$ equilibrium and then almost all the abundance in an isotopic chain is in one or two isotopes with $S_{2n} \sim 2S_n^0$ as mentioned earlier. Such isotopes are the waiting-point nuclei as the r-process must wait for their beta decay in order to produce nuclei in the next isotopic chain [19]. The special case of steady flow mentioned here is named the steady beta-flow. The yield of the waiting-point nucleus, in this case, is inversely proportional to the beta-decay rate, as is clear from Eq. (5.24).

However, the results for a steady flow calculation are valid only for certain regions of the network where nuclei are fed into the reaction network from below for a sufficiently long time. It is seen that a steady flow is indeed realized for the waiting-point nuclei between the magic neutron numbers 50 and 82. Thus with the idea of steady flows in the r-process, the peak at $A = 130$ and $A = 195$ in the SoS abundances can be explained by being connected to the really small beta decay rates of the waiting-point nuclei with N=82 and 126, the two neutron magic numbers.

5.2.2 Fission Cycling

For the r-process progenitor nuclei having $Z \geq 80$, one needs to include fission cycling as mentioned earlier. In this situation, the heaviest nucleus produced through the r-process fissions and a cyclic flow occurs involving the nucleus and its fission fragments. This, of course, has to be in an environment of high neutron abundance. We take the example of fission happening upon the beta decay of the heaviest waiting-point nucleus with Z_f protons and breaking it up into two lighter nuclei with proton numbers Z_1 and Z_2 ($Z_2 > Z_1$) in a situation where the equilibrium $(n, \gamma) \rightleftharpoons (\gamma, n)$ exists. The abundances of nuclei with $Z < Z_1$ after sufficient time are depleted to zero through neutron capture and beta decay. The yields of the nuclei involved in the fission cycling are given by Qian [19]

$$\frac{dN(Z)}{dt} = \lambda_\beta(Z-1)N(Z-1) - \lambda_\beta(Z)N(Z) , \tag{5.25}$$

for $Z_1 < Z < Z_2$ and $Z_2 \leq Z \leq Z_f$ where $N(Z) = \Sigma_A N(Z, A)$ as in Eq. (5.24). Also

$$\frac{dN(Z_1)}{dt} = \lambda_\beta(Z_f)N(Z_f) - \lambda_\beta(Z_1)N(Z_1) , \tag{5.26}$$

$$\frac{dN(Z_2)}{dt} = \lambda_\beta(Z_f)N(Z_f) + \lambda_\beta(Z_2 - 1)N(Z_2 - 1) - \lambda_\beta(Z_2)N(Z_2) . \tag{5.27}$$

Thus with the feeding of the nuclear flow with the lower proton numbers $Z = Z_1$ and $Z = Z_2$ by fission, a steady state is reached after a few fission cycles. This result helps in yield patterns to peak at $A \sim 130$ and $A \sim 195$.

5.2.3 Freeze-Out

In a realistic situation for the r-process, the temperature and the neutron density often decrease with time. One says that the r-process becomes inefficient and reaches a freeze-out when

$$N_n < v\sigma_{n,\gamma} > \tau \sim 1 , \tag{5.28}$$

where τ is the timescale of the decrease of the neutron density. The freeze-out may happen in a few stages during which the conditions required for the r-process break down. With the freeze-out, the progenitor nuclei go toward stability over a number of beta decays shown in Fig. 5.2 by dotted arrows. Specifically the nuclei with r-process peaks with magic neutron numbers change into nuclei with lower non-magic neutron numbers but with the A values remaining roughly unchanged. However, processes such as beta-delayed neutron emission can change this pattern of freeze-out significantly.

5.2.4 Conditions Needed for the r-Process

As the site for the r-process must have a very neutron-rich matter, one naturally thinks of the deep interior of massive stars going through normal core collapse supernova explosions, or explosions of some special class of supernovae like the rotational MHD-jet ones or collapsars. The other possibility is ejecta from events like the binary neutron star mergers or neutron star merging with black holes. As discussed in Chap. 2 for supernovae, the n-rich matter at high temperatures with the matter dissociated into nucleons, cool almost fully through the emission of neutrinos/antineutrinos and the dominant charged current (CC) interaction with the matter in the ejecta is given by Qian and Woosley [24]

$$\nu_e + n \rightleftharpoons p + e^-, \tag{5.29}$$

$$\bar{\nu}_e + p \rightleftharpoons n + e^+. \tag{5.30}$$

This interaction of the neutrinos and antineutrinos transfers energy from the $\nu/\bar{\nu}$s to the matter above the neutron star in the SN scenario and heat it. Consequently, this matter expands away from the proto-neutron star and develops into a mass outflow which is called the neutrino-driven wind.

For the situation with the neutrino-driven wind outside the proto-neutron star or even for the NS merger case, the conditions for the equilibrium of the Y_e after the heating by ν_e or $\bar{\nu}_e$ can be obtained analytically. The rate of change of the electron fraction can be written as

$$\frac{dY_e}{dt} = (\lambda_{\nu_e n} + \lambda_{e^+ n})(1 - Y_e) - (\lambda_{\bar{\nu}_e p} + \lambda_{e^- p})Y_e = \lambda_1 - \lambda_2 Y_e, \tag{5.31}$$

where $\lambda_{\nu_e n}$, $\lambda_{e^+ n}$ are the rates of the reactions that destroy neutrons and create protons in Eqs. (5.29) and (5.30). Their sum is written as λ_1. Also $\lambda_{\bar{\nu}_e p}$ and $\lambda_{e^- p}$ stand for the rates for the reactions that destroy protons and create neutrons [24]. λ_2 stands for $\lambda_1 + \lambda_{\bar{\nu}_e p} + \lambda_{e^- p}$. Then for $\frac{dY_e}{dt} = 0$ one gets the equilibrium value of Y_e, i.e. $Y_{e,eq.}$.

$$Y_{e,eq.} = \frac{\lambda_1}{\lambda_2} = \frac{1}{1 + \frac{\lambda_{\bar{\nu}_e p} + \lambda_{e^- p}}{\lambda_{\nu_e n} + \lambda_{e^+ n}}}. \tag{5.32}$$

Using the heating terms for the matter in the neutrino-driven wind which essentially comes from the neutrino/antineutrino capture at matter temperatures around 1 MeV [19], one can drop $\lambda_{e^- p}$ and $\lambda_{e^+ n}$ and write using the expressions for the neutrino and antineutrino heating involving their luminosities [1]

$$Y_{e,eq.} = [1 + \frac{L_{\bar{\nu}_e} W_{\bar{\nu}_e} \epsilon_{\bar{\nu}_e} - 2\Delta + \Delta^2 / < E_{\bar{\nu}_e} >}{L_{\nu_e} W_{\nu_e} \epsilon_{\nu_e} + 2\Delta + \Delta^2 / < E_{\nu_e} >}]^{-1}. \tag{5.33}$$

Here L_{ν_e} and $L_{\bar{\nu}_e}$ are the neutrino and antineutrino luminosities, $\epsilon_{\nu_e} = < E_\nu^2 > / < E_\nu >$ stands for the ratio between the second moment of the neutrino spectrum and the average neutrino energy. One also has a similar expression for $\epsilon_{\bar{\nu}_e}$. The quantity Δ stands for the neutron-proton mass difference which is 1.2933 MeV. The weak magnetism corrections to the neutrino and antineutrino capture cross sections are given by Horowitz [25] $W_\nu = 1 + 1.01 < E_\nu > /(m_u c^2)$ and $W_{\bar{\nu}} = 1 - 7.22 < E_{\bar{\nu}} > /(m_u c^2)$ where m_u is the mass of the nucleon.

If the matter is exposed to the neutrinos long enough so that the reactions reach an equilibrium, then the matter becomes neutron-rich only when the following condition is observed

$$\epsilon_{\bar{\nu}_e} - \epsilon_{\nu_e} > 4\Delta - [\frac{L_{\bar{\nu}_e} W_{\bar{\nu}_e}}{L_{\nu_e} W_{\nu_e}} - 1](\epsilon_{\bar{\nu}_e} - 2\Delta) . \qquad (5.34)$$

For the SN scenario, with the $\bar{\nu}_e$ spectrum slightly hotter than the ν_e one, the luminosities of both flavors are similar and then the average energies as given in Eq. (5.34) should differ by more than $4\Delta \sim 5.2$ MeV. But the realistic simulations do not show such a large difference in average energies. However, for the neutron star mergers, the situation is different. There one starts with cold very n-rich matter and during the merger it gets heated to high temperature favoring the formation of $e^- e^+$ pairs and through protonization of the matter, the new equilibrium is reached in timescales of a few times 0.1 s. During this the luminosities and average energies of $\bar{\nu}_e$ are much larger than those of ν_e, which reduces the required energy difference, as can be seen from Eq. (5.34). Thus the neutrino interactions in mergers make the late ejecta enough neutron-rich to cause a weak r-process. But the early dynamic ejecta, coming out of the spiral arms after the collision, stay very n-rich and gives rise to a strong r-process [1].

Another effect that disfavors the r-process nucleosynthesis in supernovae is the material goes through an alpha-rich freeze-out due to high entropies and moderate electron fractions. This means that under the strong neutrino fluxes, the phase of alpha formation [26] pushes the matter toward $Y_e \sim 0.5$, prohibiting the r-process. But in the case of NS mergers, due to moderate entropies, for $Y_e \leq 0.45$ the alpha-effect is absent.

The effect of neutrino reactions on Y_e has also to take into account the oscillations of active flavors in presence of matter, the possible active-sterile transformations, as implied by the reactor [27] and Gallium anomalies [28], as well as the effect of neutrino self-interaction giving rise to the collective neutrino oscillations. The collective oscillations were discussed in Chap. 2 and present studies indicate that these oscillations may indeed have effects on the Y_e and the r-process nucleosynthesis [29–31].

Following Cowan et al. [1], we list in the next section all the possible sites, first in the collapse of massive stars and then in neutron star mergers.

5.2.5   The Collapse of Massive Stars as Site for the r-Process

In Chap. 2, we discussed the scenario for the collapse of massive stars followed by the bouncing of matter and creation of a shock wave that moves outward. But as pointed out there, except for lower mass stars with stellar masses $9 - 10\ M_\odot$ one does not get a prompt explosion. The delayed heating of matter ahead of the shock by the huge flux of neutrinos and antineutrinos streaming out, may be the way one gets explosions and Chap. 2 gives the detailed results of the present hydrodynamical simulations. The occurrence of the r-process has been proposed for different stages of the collapse of large stars. One of them is during the heating of matter by the neutrino-driven wind in the delayed explosions and early results showed it to be a promising site. But recent studies indicate that with realistic mass models [32], one needs a superposition of high entropies, like 120–280 k_B for $Y_e = 0.45$ to produce abundances close to the SoS values [33]. But such high entropies are not reached in present simulations. Also, as mentioned in Sect. 2.9, the similarity of ν_e and $\bar{\nu}_e$ spectra prohibits the reactions for the r-process. However, it can still be a weak r-process.

A class of supernovae named electron capture supernovae with progenitor star masses 9–10 $M_\odot$ with a degenerate O-Ne-Mg core is predicted to implode initially after its core runs out of nuclear fuel and eventually causing a prompt explosion. The simulation results of such a case done in one dimension for a star of mass 9 $M_\odot$ is shown in Fig. 5.5. This may again result in a weak r-process [33–35] and thus producing nuclei up to Eu, but not near or above the r-process peak around Pt [36]. Also for core collapse of stars of masses 11–15 $M_\odot$ with low metallicities like $[Fe/II] \leq 3$, the neutrons produced through the reaction $^4He(\bar{\nu}_e, e^+n)^3He$ in the He-shell can be absorbed by nuclei with mass numbers as large as 200.

The possibility of obtaining quark deconfinement during neutron star formation was discussed in Chap. 3. Though it is still speculation but the assumption that

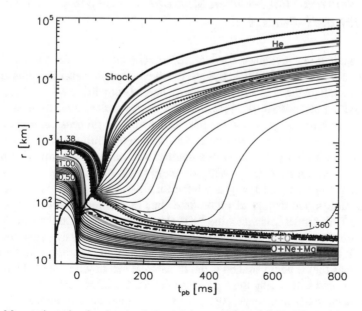

Fig. 5.5 Mass trajectories for the simulation of the explosion of O-Ne-Mg core star of mass 9 $M_\odot$. The thick solid line starting at time zero gives the shock position and the neutrinospheres (ν_e : thick solid, $\bar{\nu}_e$: thick dashed, ν_μ, $\bar{\nu}_\mu$, ν_τ, $\bar{\nu}_\tau$:thick dash-dotted) are also indicated. Credit: F. S. Kitaura et al., Astron. Astrophys. **450**, 345 (2006) reproduced with permission ©ESO

quark-hadron phase transition takes place at an appropriate stage in the density-temperature conditions, may make the proto-neutron star go through a collapse again. If the second collapse does not take the matter to black holes but halts, the r-process is possible in the innermost ejecta in some favorable conditions.

The core collapse of stars with fast rotation and strong magnetic field which results in supernova explosions gives rise to neutron stars with very high magnetic fields $\sim 10^{15}$ G and is known as magneto-rotational supernovae or MHD-SNe [1]. The MHD calculations assuming axis symmetry showed the r-process synthesis [37] and later calculations were done to check whether general relativistic 3-dimensional simulations also get this effect. The results are summarized in Cowan et al. [1] concluding that this mechanism can produce magnetar-strength magnetic fields and can cause magneto-rotationally powered explosions. This can happen for smaller initial magnetic fields and that may be a possibility how most magnetars are generated. The discussions in Cowan et al. [1] also include conditions for prompt polar jet-like ejection of n-rich matter as well as delayed ejectas which go through more interaction with neutrinos with the results of having higher Y_e and weakened strength of the r-process.

There are also suggestions for the r-process in the very energetic supernovae named hypernovae as well as in collapsars [1].

5.2.6 Neutron Star–Neutron Star/Black Hole Merger as Site for the r-Process

There is now direct evidence for NS-NS merger as a site for the r-process in the observation of GW170818. But as the next section describes this event in detail, we shall make the discussion short here. From the days of discovery of Hulse–Taylor pulsar, PSR1913+16, the first binary pulsar system seen in 1974, the interest in binaries grew, and the possibility of the r-process taking place there was discussed. Around the same time, Lattimer and Schramm [38] proposed the coalescence of a binary system with a neutron star and a black hole that provides a source of neutron-rich ejecta. With the early work of Li and Paczyński [39] in 1998, Metzger et al. in 2010 [40] calculated for NS-NS merger the late-time heating from the radioactive decays of nuclei and thus connected NS-NS merger to the r-process. This model known as the "Kilonova model" is described in Sect. 5.6. After the observational support in 2017, a lot of work is presently being done in the framework of the kilonova model to make its predictions more robust. One of the specific areas that is being studied is the dynamical ejecta with the cold and the hot components and calculating the r-process abundances after some decay time, as a function of the mass number A. Again the neutron capture and the fission rates are important quantities in determining the production of the r-process elements. In this scenario too, one gets a neutrino wind but its structure is different from the CCSN case. The electron type antineutrino luminosity is much larger than the electron neutrino luminosity and the difference of their average energies is also large. Most calculations, as yet, are based on the neutrino leakage scheme and better treatment of the neutrino diffusion including possible realistic neutrino flavor conversion in future is envisaged.

It should be mentioned here that in 2020 a NS-BH merger was observed through the detection of the emitted gravitational waves by the LIGO Scientific Collaboration (LSC), the Virgo collaboration, and Kamioka Gravitational Wave Detector (KAGRA) project, but the associated electromagnetic radiation of the event was not seen.

5.3 Inputs for Nuclear Modelling of the r-Process

5.3.1 Nuclear Masses/Binding Energies

We briefly present, following Cowan et al. [1], the current status of the models for prediction of masses of all possible nuclei and discuss how accurate they are compared to the nuclei whose binding energies are known experimentally. There are efforts spread over the last few decades to build successful models for the masses. Some of them are:

(a) microscopic-macroscopic models like (1) Finite-Range Droplet Model (FRDM-1992) [32] (2) Extended Thomas-Fermi model with Strutinsky Integral approach (ETFSI) given in Aboussir et al. [41] (3) the Extended Bethe-Weizsäcker and Weizsäcker-Skyrme mass models [42] along with the model of Liu et al. [43] named WS3.
(b) a parameterized model using the averaged mean field taken from the shell model and with the addition of Coulomb, pairing and symmetry energies named DZ10 and DZ31 [44].
(c) models based on the non-relativistic fields [45] given in Goriely et al. [46] named HFB-21 as well as their relativistic versions [47].

Table 5.1 gives the root mean square deviation of the predicted masses from the experimental values. The experimental values are from the 2003 compilation of Audi et al. [48] named AME-2003 and the 2012 values presented in Wang et al. [49], named AME-2012. For the understanding of the results of given in Table 5.1 one needs to point out some details. The AME-2012 set added 219 new observed masses to the set off AME-2003. The rms deviations of the "full" AME-2003 is seen to vary from 336 to 655 keV whereas the "full" larger AME-2012 dataset has a range from 345 to 666 keV for the five different models. Out of these WS3 is the most successful for both sets. But when one looks at only the 219 "new" set of experimental values, the predictability goes down somewhat, with the deviation now ranging from 424 to 880 keV. This is because the more recent experiments are observing the binding of the nuclei far away from the stability line mostly in the r-process range. So fitting the

Table 5.1 The rms deviation of the predicted masses in keV by the FRDM-1992, HFB-21, WS3, DZ10 and DZ31 models from the experimental values of AME-2003 and AME-2012. The columns labelled "full" considers all the masses present while the column "new" includes only the masses found in AME-2012 but not in AME-2003 (taken from Cowan et al. [1])

Model	AME-2003 (full)	AME-2012 (new)	AME-2012 (full)
FRDM-1992	655	765	666
HFB-21	576	646	584
WS3	336	424	345
DZ10	551	880	588
DZ31	363	665	400

parameters of a model to all observed masses and then seeing how the predictions turn out to be for the very neutron-rich nuclei is a challenge for the models. Finally, it should be mentioned that though the different mass models do not differ much in their predictions, they are not equally successful in the r-process simulations, particularly in the transitional regions of neutron numbers around 90 and 130. But overall with a better understanding of the physics of the nuclei close to the drip line, the models are improving and pushing the predicted abundances closer, in general, to the measured values.

5.3.2   Beta Decay Rates

We remind us that once reactions involving the capture of charged particles like protons stop due to strong Coulomb repulsion, the way nuclei up to the uranium/thorium get produced is through the capture of one or multiple neutrons in neutron-rich environments, followed by beta decays. So to decide whether it is possible to reach the uranium end in the timescale available, one needs to know the beta decay rates of the many decays that are involved. Also, in the r-process scenario when repeated neutron capture takes place, the Q-value of the very neutron-rich nuclei at the end are quite large, often having values like 15–20 MeV. The allowed beta decay strength goes as Q^5, neglecting the Coulomb distortion factor (see Eq. (2.13) in Chap. 2) and the energy dependence of the Gamow-Teller strengths (the forbidden decays have even a higher dependence on Q value). Then the corresponding rates of these nuclei are very large and consequently the half-lives very short. However in reality the rates depend sensitively on the Gamow-Teller strength energy distribution and so one needs to evaluate the matrix elements of the Gamow-Teller operator between the decaying state of the mother and the ground and excited states of the daughter nucleus. Also some of the high-lying states of the daughter are above the neutron emission threshold and hence have neutrons emitted spontaneously, a process known as beta-delayed neutron emission. The neutron emission threshold is about 2–3 MeV for most the r-process nuclei.

Particularly important are the beta decays of the r-process nuclei with neutron number $N = N_{magic}$, N_{magic} being the neutron magic number. This is because for nuclei with a magic number of neutrons, its extra binding, due to the pairing and the shell closure reduces the neutron separation energy of the nuclei with neutron number equal to $N_{magic} + 1$ substantially, hindering the matter flow. Also reduced Q-value of the nuclei with magic neutron number leads to longer lifetimes.

In the early days some very important experiments to find the half-lives of the r-process n-rich waiting-point nuclei were done for the ones with N=50, i.e. ^{80}Zn and ^{79}Cu; also for two other with N=82, i.e. ^{129}Ag and ^{130}Cd [50–52]. These data play crucial roles in constraining the nuclear models. Also studies of the Gamow-Teller (GT) allowed strength distribution show a long tail at high excitation energies beyond the GT giant resonance and only a small fraction of the total GT strength is reached by the β^- Q-value. In β^- decay both the Fermi and Gamow-Teller strength centroids lie above the Q-value. For β^+ decay, the Fermi transition is not allowed by the isospin selection rule and can come only through the very small isospin mixing. Also for β^+, though the GT centroid lies within the Q-value, the total β^+ strengths are much smaller compared to the total β^- strengths. We mentioned this in Chap. 2, but to emphasize we discuss this point in more detail. Using the commutation relations of the spin and isospin operators involved in the expressions of Fermi and GT β^- and β^+ operators, Ikeda [53] obtained the sum rules, i.e. total strengths summed over all possible excitation energies of daughter nuclei, as

$$S_{\beta-}^{F} - S_{\beta+}^{F} = (N - Z) ,$$
(5.35)

$$S_{\beta-}^{GT} - S_{\beta+}^{GT} = 3 (N - Z) .$$
(5.36)

These are known as Ikeda sum rules and are exact results. The half-lives, however, depend sensitively on the fraction and distribution of the GT strength in the low-lying energy region, and hence different models show large differences in their predicted values. As the models need to be used for a large number of nuclei, they are based on semiempirical global models or quasiparticle random phase approximation (QRPA). FRDM global model [32] and ETFSI [41] are the examples of the ones that are used.

In Fig. 5.6 the predictions of different models, ETFSI, FRDM, HFB and Shell Model (SM), for nuclei with N=82, are compared to experimentally known values [52]. Later experimental data of RIKEN for N=82 and Z=45 and Z=46 became available and they again showed agreement with SM. Such predictions are available for N=126 also [13, 54]. The SM calculations are possible for the magic nuclei with N=82 and 126, as the dimensions of the model spaces involved in the SM for them are not very large. In contrast, for the heavy nuclei with N and Z in between magic numbers where active nucleons occupy a number of valence orbits, complete SM calculations involve diagonalization of the interaction Hamiltonian in spaces of exceedingly large dimensions and as a result only truncated calculations are possible. Figure 5.6 clearly shows good agreement that SM results have with

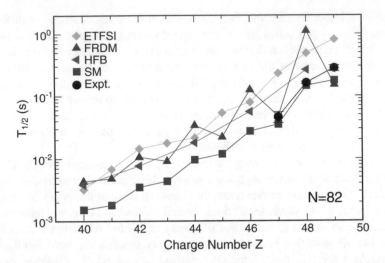

Fig. 5.6 The half-lives of the r-process waiting-point nuclei with neutron number N=82 obtained by Shell Model (SM) and other theoretical models Reprinted figure with permission from [K. Langanke, G. Martinez-Pinedo, Rev. Mod. Phys. **75**, 819 (2003)] ©2003 by the American Physical Society (APS). https://doi.org/10.1103/RevModPhys.75.819

experimental values and this agreement is better than those by other models. Another example is the nucleus ^{78}Ni for which the SM predicted a half-life of 127 ms [13] which was shorter than the predictions of others. But then it later agreed with the experimental value of $110(^{+}_{-}40)$ ms obtained by Hosmer et al. [55]. So overall the most reliable method for the calculation of the Gamow-Teller matrix elements or strengths as a function of the daughter nucleus energy is the Shell Model, particularly for the spherical or near-spherical nuclei.

Finally the charged-current (ν_e, e^-) reactions in the region of high flux of ν_e may also affect the r-process. They also involve the allowed GT strength distribution for their reaction cross sections. But as discussed earlier in Sect. 5.2.4, the neutrino-driven wind scenario is not a favored site for the r-process.

5.3.3   Neutron Capture Rates

At temperatures higher than about 10^9 K during the r-process, the neutron capture reactions and the photo-dissociation reactions remain in equilibrium. But when the temperature falls below 10^9 K, one needs to know the neutron capture rates accurately for determining the r-process path. The calculations for these rates are done using the statistical model. Here the density of states at the relevant energies is considered large enough and the Hauser-Feshbach theory used is based mainly on three quantities- the nuclear level density, the gamma strength functions of different multipolarities and the light particle potentials. As parity and angular momentum are conserved, one needs to know the level densities with fixed angular momentum and parity. Detailed results in very large spaces are available now for such level densities using the version of Shell Model known as Shell Model Monte Carlo (SMMC) approach [56, 57] and its later versions are presented in the references [58, 59] and [60]. In Mocelj et al. [61] this method has been used for a large set of the r-process nuclei using a temperature dependent pairing parameter suggested from SMMC studies and their effects on the reactions for astrophysics were looked into by Loens et al. [62]. The statistical package NON-SMOKER and SMARAGD, developed by Rauscher, [63–65] have the improved treatment of level density incorporated. Goriely and collaborators used a combinatorial approach based on the framework of HFB and developed a package for calculating the r-process neutron capture rates named as Brussels Nuclear Library for Astrophysics Applications (BRUSLIB, http://www.astro.ulb.ac.be/bruslib) [66]. The different gamma strength functions involving parameters that are adjusted to photo-dissociation results for E1 and electron scattering data for M1 transitions [67], are useful. The transmission coefficients needed in the statistical model for astrophysical reaction rates [63, 67] are based on global optical potentials and both for neutrons and protons such potentials are available. For details, we refer to Cowan et al. [1].

We should also mention that fission of the relevant heavy nuclei are also important inputs, especially for the merger of neutron stars as discussed in Chap. 4.

5.4 Electromagnetic Counterpart of GW170817 and Ejected Matter in BNS Merger

The discovery of the first binary neutron star (BNS) merger GW170817 has been a huge boon in the field of nuclear astrophysics research. This event was not only detected in gravitational waves, but it was also observed across the electromagnetic (EM) spectrum [68, 69]. A short Gamma Ray Burst (sGRB) was recorded by the Fermi-GBM and INTEGRAL telescopes 1.7 s later following the BNS merger in GW170817 [69]. This led to an immediate follow-up campaign for EM observations using a large number of ground based telescopes around the world as well as space based telescopes [69]. Indeed the first EM counterpart, referred to as AT 2017gfo, associated with a BNS merger event was observed almost 11 hours after the merger event GW170817 [70, 71]. It was noted from Ultra-violet (UV)/Optical observations that the early blue emission faded away in just 2 days. The detection of redder optical and near-Infrared (IR) emissions following the blue emission lasted for 2–3 weeks. Later X-ray and radio emissions were observed from the transient on 9th and 16th days, respectively. The spectra of early emissions from the EM counterpart were described by a blackbody having a bolometric luminosity $\sim 5 \times 10^{41}$ erg s^{-1}. These features of AT 2017gfo are consistent with the predictions of a Kilonova (KN) model [40].

It may be argued that a KN shines due to the decays of the r-process radioactive nuclei synthesized in the ejected material in a BNS merger. These observations in UV/Optical/IR bands point to the presence of two separate components in the ejected material. The blue emission denotes the matter with a large electron fraction or neutron-poor matter in which the light r-process elements are synthesized. On the other, the red/infrared emission originates from the highly neutron-rich ejected matter that is responsible for the synthesis of the r-process heavy nuclei.

There are two distinct ways for the production of the ejected matter according to the investigations in general relativistic simulations of BNS mergers [72, 73]. The matter is dynamically ejected in the polar direction as well as on the equatorial plane during the merger that happens on the time scale of a few tens of milliseconds. In this scenario, the faster moving material in the polar direction is liberated at the interface of two colliding neutron stars due to the shock heating. This matter further interacts with neutrinos coming out of the hypermassive neutron star (HMNS) making this ejecta a neutron-poor material having low opacity. The electron fraction in this case could be >0.3. This matter represents the blue KN. This component may not be present if a prompt black hole is formed in the merger. On the other hand, tidal forces strip off materials from the surfaces of two approaching compact stars and throw this on the equatorial plane. It is a neutron-rich matter with low electron fraction <0.2 and endowed with a high opacity due to the presence of heavy elements synthesized in the r-process. This neutron-rich ejecta contributes to the red kilonova with longer lasting redder emission. Velocities of dynamical ejecta lie in the range $\sim 0.1 - 0.3c$, where c is the speed of light. Furthermore, additional contributions to the blue and red kilonovae might come from the ejected matter from the disk around the merger remnant. The mass loss from the disk is driven by neutrino heating at early times. This is similar to the neutrino driven proto-neutron star winds in the CCSN. This material has a broad electron fraction $\sim 0.1 - 0.4$. Figure 5.7 depicts all these processes of the matter ejection in a BNS merger. The blue KN is made of an ejected mass of $\sim 10^{-2}$ $M_\odot$ whereas the amount of red KN material is $\sim 5 \times 10^{-2}$ $M_\odot$. The total kinetic energy of blue and red ejecta is $\sim 10^{51}$ ergs [74].

We mention here the work of Perego, Thielemann and Cescutti [75] which reviews the conditions of matter expelled in the mergers, for both BNS and BBHNS and describes how the r-process takes place there.

5.5   Decompression of Ejected Neutron-Rich Matter in Lattimer and Schramm Model

The discovery of the Hulse–Taylor pulsar in the double neutron star system PSR 1913+16 in 1974 [76] provided important clues for the r-process nucleosynthesis sites in binary compact star mergers. It was estimated that the BNS merger in PSR 1913+16 would occur in ~ 300 million years [77]. Lattimer and Schramm [38, 78, 79] and Symbalisty and Schramm [77] proposed that the neutrino-rich

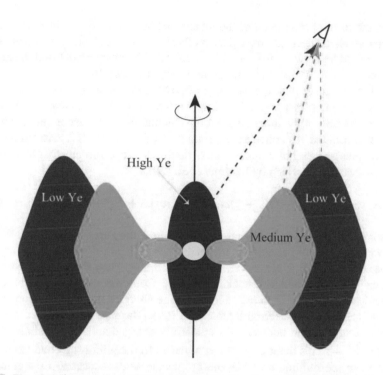

Fig. 5.7 The ejected matter in a BNS merger along the polar direction and equatorial plane are shown here. The merger remnant is denoted by the yellow sphere at the center. The dark yellow regions represent the accretion disk around the remnant. The (deep) blue and red regions are the blue and red kilonovae. Reprinted figure with permission from [M. Shibata et al., Phys. Rev. **96**, 123012 (2017)] ©2017 by the American Physical Society (APS). https://doi.org/10.1103/PhysRevD.96.123012

material might be expelled either in the tidal disruption of the neutron star by a black hole or in BNS mergers. They looked into the question of what kind of nuclei would be formed in the ejected matter due to the breakup of a neutron star by the tidal force of a black hole [78].

We have already noted in Chap. 3 that the compositions of neutron star matter changes as the density increases from the surface to the center of the compact star. Broadly the outer crust is made of nuclei and the nuclear clusters are immersed in a gas of free neutrons in the inner crust in the background of a noninteracting electron gas and the neutron star core above the saturation density is composed of β-equilibrated and charge neutral matter of neutrons, protons, other exotic forms of matter (hyperons, antikaon condensate and quarks) along with electrons. The ejected matter from the neutron star should undergo a decompression as envisaged by Lattimer et al. [79]. This happens due to the expansion of the matter over a free-fall time scale $\sim 446\,\rho^{-1/2}$ s, where ρ is the mass density of the ejected matter. Consequently, the density decreases as the ejected matter above the saturation density expands in equilibrium. As the density falls below the saturation density, it takes the form of the matter of the inner crust but out of β-equilibrium [78, 79]. Lattimer et al. considered the nucleosynthesis in the cold ejected matter whose density would lie between the saturation density, $\sim 3 \times 10^{14}$ g/cm^3, and the neutron drip density $\sim 4 \times 10^{11}$ g/cm^3 relevant to the inner crust [79].

The initial state of the expansion is given by the pre-breakup ground state matter in equilibrium. Lattimer and collaborators adopted the Baym, Bethe, and Pethick (BBP) model [80] of the inner crust for the calculation of the decompressed matter. Nuclei are arranged in a lattice and surrounded by free neutrons at zero temperature. The total energy (E_N) of a nucleus of N neutrons and Z protons that includes the lattice contribution is given by Lattimer et al. [79],

$$E_N(A, y, k, R) = [W + ((1 - y)m_n + ym_p)c^2]A + E_C + E_S + E_{shell}, \quad (5.37)$$

where $A = N + Z$, proton fraction $y = Z/A$ and k the Fermi momentum. W is the energy per nucleon of the uniform nuclear matter, E_C and E_S are the Coulomb and surface energy, respectively and E_{shell} is the semiempirical shell energy correction as given by Myers and Swiatecki [81]. The Coulomb energy contains all the contributions, i.e. nuclear, lattice, and exchange energies of proton-proton, proton-electron, and electron-electron interactions [80]. The nuclear Coulomb energy is obtained assuming the nucleus as a uniformly charged sphere of radius "R" and given by $\frac{3}{5}\frac{Z^2e^2}{R}$. This inner crust matter satisfies four equilibrium conditions which follow from the minimum of the energy density with respect to its argument at a given baryon number density (n_B) [79, 80]. Those are the optimum number of nucleons in a nucleus as found from $\frac{\partial(E_N/A)}{\partial A}|_{A,y,k,R,n_B} = 0$, the nucleus in β-equilibrium, and the equality of neutron chemical potentials and pressure in the matter inside the nucleus and the neutron gas.

With this initial state of the equilibrium matter, Lattimer and collaborators explored the expansion of a unit cell of matter containing one nucleus. As the density falls due to the expansion, the pressure of the neutron gas outside the nucleus is reduced. As a result, neutrons come out of the nucleus. In this situation, the matter is out of equilibrium as there would be no β-decays because the β-decay timescale is longer than the expansion timescale. But the two timescales become comparable when the density drops to 10^{11} g/cm^3 and this facilitates the β-decay. The temperature of the matter might reach a value of $\sim 10^9$ K due to the energy release in β-decays. Free neutrons in the gas phase could be captured once the β-decay increases Z inside the nucleus. The nucleus grows in neutron and proton numbers until it becomes unstable to fission. Fissions of nuclei release more energy and increase the temperature of the system further to 10^{10} K. This might lead to the nuclear statistical equilibrium in the system. It is to be noted that there would be no free neutrons when the density drops below the neutron drip point due to the expansion. All free neutrons are quickly captured by nuclei before the expansion brings down the density below the drip point. Lattimer and collaborators showed that the path of synthesized nuclei in the Z-N plane appeared on the neutron-rich side of the β-stability line [79]. This whole process of nucleosynthesis in the ejected matter resembles very much the r-process as discussed in Sect. 5.2.

5.6 Kilonova Model

Kilonovae are optical transients created in BNS or binary black hole neutron star (BBHNS) mergers [40, 82]. The naming of a transient as kilonova (KN) was justified by B. Metzger and collaborators because the luminosity of the transient peaks at a value that is $\sim 10^3$ times higher than that of a typical nova [40]. S. Kulkarni discussed similar transients powered by ^{56}Ni and free neutron decays and named those macronovae [82]. It is worth mentioning here that the ^{56}Ni cannot be synthesized in the neutron-rich ejecta of a BNS merger. Radioactive decays of the r-process nuclei formed in the ejected neutron-rich material in those mergers power a KN. The first observation of a KN using the Hubble Space Telescope was reported by Tanvir et al. [83] and Berger et al. [84]. They also suggested that kilonovae would be the EM counterparts of gravitational wave signals in compact object mergers and important as the r-process nucleosynthesis sites. A KN followed by the gravitational wave signals in the BNS merger GW170817 was detected and investigated at length for the first time. The EM transients in binary compact object mergers evolve faster than supernovae. Here we highlight the salient features of a KN and discuss how this enriches our knowledge about the r-process nucleosynthesis.

Li and Paczyński developed a simple model of such a transient and constructed light curves using this model [39]. This model assumed that neutron-rich radioactive nuclei produced in the expanding ejected material of BNS and BBHNS mergers provided a long-term source of heat. They obtained the luminosity (L) from an overall heat balance relation [39],

$$L \simeq (\dot{\varepsilon} - T\frac{dS}{dt})M, \qquad (5.38)$$

where the first term gave the radioactive heat generation per gram of the matter per second,

$$\dot{\varepsilon} = \frac{fc^2}{t}, \qquad (5.39)$$

and the second term represented the heat loss in the adiabatic expansion. Here f was the fraction of the rest mass energy and treated as a free parameter, and M is the total mass of the ejected matter. Based on this model, Li and Paczyński obtained the light curves of the KN numerically and analytically. It was noted that the transient could reach a peak luminosity as large as $\sim 10^{44}$ erg/s for $f = 10^{-3}$ and the luminous phase might last for ~ 1 day [39]. This short time scale of the peak luminosity of the merger transient is in stark contrast with that of a typical supernova $t_{peak} \sim$ weeks.

Later it was realized that the values of $f = 10^{-5} - 10^{-3}$ resulted in high heating rates leading to much brighter kilonovae than typical supernovae [85].

The first realistic modeling of a KN taking into account the radioactive heating using the nuclear reaction network in a self-consistent fashion and with inputs from the dynamical ejecta trajectories of Rosswog et al. was performed by B. Metzger et al. [40, 86]. In this connection, they used a dynamical reaction network including neutron captures, β and α decays, fissions, and photo-disintegrations of nuclei. Neutron capture cross sections and nuclear masses for the r-process in their calculations were estimated based on two nuclear mass models—FRDM and the quenched version of ETFSI. Initial values of density, temperature (6×10^9 K), electron fraction, and mass, and atomic numbers of seed nuclei were provided for a given calculation. In the r-process heating, radioactive decays of nuclei synthesized in the r-process inject the heat in the ejecta through β-decays, α-decays, and fissions. Further, the decays of free neutrons might be another source of the heating. Generally, all neutrons in the ejecta are captured during the r-process nucleosynthesis in a timescale ~ 1 s. But a small fraction of neutrons was found not to be captured by nuclei in BNS merger simulations [40, 85]. These decays of free neutrons can enhance the luminosity of a KN appreciably during the early hours. All the heat generated in the r-process was added self-consistently to the entropy of the fluid in this work. It was noted that only a fraction of the total energy released in nuclear reactions that would thermalize with the plasma, might power the EM radiation. For a fixed energy release, the thermalization efficiency in fission, α-decay, and β-decay happens in descending order. Finally, this calculation was used to model the light curve and color evolution. The main findings of the KN model of B. Metzger et al. are the following.

The radioactive heating on the time scales corresponding to the peak time of the luminosity $\sim$ hours-days stemmed from the β-decays and fissions of the r-process nuclei produced at early times. This result was practically insensitive to the electron fraction. The net heating rate followed a power law $\dot{Q} \propto t^{-\Gamma}$, where $\Gamma = 1.1 - 1.4$ for $t \sim$ hours-days. This power law behavior was found to be similar to that of Eq. (5.39) in Li and Paczyński [39]. Metzger and collaborators predicted that the peak bolometric and V-band luminosities of the transient $\sim 10^{42}$ ergs/s and $\sim 3 \times 10^{41}$ ergs/s might happen at a time $t \sim 1$ day for the ejecta mass $\sim 10^{-2} - 10^{-3}$ $M_\odot$. The value of the free parameter $f \sim 3 \times 10^{-6}$ was obtained by fitting the results of Metzger et al. with the model of Li and Paczyńsk [40].

Figure 5.8 shows the bolometric luminosity of the EM counterpart of GW170817 as a function of time since the BNS merger. Data of EM observation [88] are denoted by squares and compared with the theoretical KN model of Metzger et al. [40] represented by the solid line with a power law decay $\propto t^{-1/3}$. It is observed that the bolometric luminosity peaked early and decayed thereafter. The decay of the luminosity followed the power law with $\Gamma \sim 1.3$ as predicted by Metzger et al. [40]. The estimated total mass ejected in the merger was $\sim 0.05 - 0.07$ $M_\odot$. This shows an excellent agreement between the observations and the KN model and is a confirmation of the idea that the KN is powered by the radioactive decays of

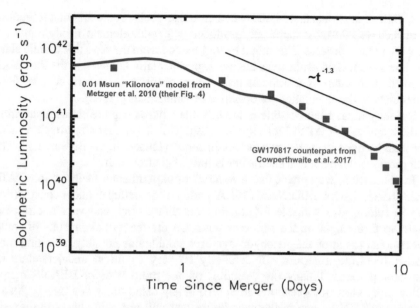

Fig. 5.8 Bolometric luminosity of a KN is plotted as a function of time since the BNS merger. Blue squares denote the data from the observations of the transient created in G170817. The solid line represents the theoretical KN model of Metzger et al. [40]. Reprinted from Ref. [87], ©2019 with permission from Elsevier

the r-process nuclei produced in the ejected neutron-rich material following a BNS merger.

5.7   Heavy Element Synthesis in Neutron-Rich Matter Ejected in GW170817

The detection of the EM counterpart AT 2017gfo associated with the BNS merger GW170817 confirmed the presence of a KN that was powered by the decays of the r-process radioactive nuclei formed in the neutron-rich ejecta. The observations of AT 2017gfo indicated that the BNS merger gave birth to a KN made of two components of the ejecta [71]. The blue component of the KN was the result of the synthesis of the r-process nuclei having a mass number $A \leq 140$ and had low opacity $\kappa \sim 1$ cm^2/g. This component in the light curve was bright and disappeared in a few days. The red component of the KN comprised of the r-process heavy nuclei with $A \geq 140$. It had large opacity $\kappa \sim 10$ cm^2/g and was visible in the infrared light curve of AT 2017gfo for a longer duration of $\sim$ weeks.

The infrared spectroscopy of AT 2017gfo led to the detection of spectral peaks in the KN light curve. These spectral features were explored using theoretical models including lanthanides and found to be in agreement with the theoretical predictions

[71]. Still it is a debatable issue whether those peaks could be unambiguously associated with the most abundantly produced lanthanide element, neodymium with $Z = 60$, in the r-process. This might be understood from the facts that atomic states and line strengths of lanthanides are not experimentally known and the theoretical calculation of those is a difficult task. As a result, there would be significant uncertainty in the calculations of opacities of lanthanides [71, 85].

Recently it has been possible to identify the neutron capture element strontium in the reanalysis of AT 2017gfo spectra [89]. This is the first confirmation of the r-process element in the neutron-rich environment created in the BNS merger. This also proves that the neutron star matter is indeed neutron-rich.

In conclusion, we remark that a number of experimental facilities like FAIR, FRIB, HIAF, RAON, RIKEN and SPIRAL, already operating or under construction, will in future, give valuable information on neutron-rich and very neutron-rich nuclei to throw light on the r-process path. On the theory side, results of multi-dimensional radiation magnetohydrodynamic simulations possibly including more realistic neutrino transport will definitely be very useful in understanding the r-process properly. Finally the detection of one more BNS or BBHNS merger followed by electromagnetic radiation will surely settle many key issues. Also if a galactic CCSN is seen in the near future that will not only clarify the physics of the late stages of the explosion but may let us know whether it is a site for the production of heavy elements by the r-process.

References

1. Cowan, J.J., Sneden, C., Lawler, J.E., et al.: Origin of the heaviest elements: the rapid neutron-capture process. Rev. Mod. Phys. **93**(1), 015002 (2021)
2. Burbidge, E.M., Burbidge, G.R., Fowler, W.A., Hoyle, F.: Synthesis of the elements in stars. Rev. Mod. Phys. **29**(4), 547–650 (1957)
3. Cameron, A.G.W.: Nuclear reactions in stars and nucleogenesis. Pub. Astr. Soc. Pac. **69**, 201 (1957)
4. Lodders, K., Palme, H., Gail, H.P.: Abundances of the elements in the Solar System. In: Trümper, J.E. (ed.) Landolt-Börnstein - Group VI Astronomy and Astrophysics 4B (Solar System), pp. 560–598. Springer, Berlin (2009)
5. Arnould, M., Goriely, S., Takahashi, K.: The r-process of stellar nucleosynthesis: astrophysics and nuclear physics. Phys. Rep. **450**(4–6), 97–213 (2007)
6. Clayton, D.D.: Principles of Stellar Evolution and Nucleosynthesis. The University of Chicago Press, Chicago (1983)
7. Iliadis, C.: Nuclear Physics of Stars. Wiley, Hoboken (2007)
8. Rolfs, C.E., Rodney, W.S.: Cauldrons in the Cosmos. The University of Chicago Press, Chicago (1988)
9. Clayton, D.D., Fowler, W.A., Hull, T.E., Zimmerman, B.A.: Neutron capture chains in heavy element synthesis. Ann. Phys. **12**(3), 331–408 (1961)
10. Käppeler, F., Beer, H., Wisshak, K.: S-process nucleosynthesis- nuclear physics and the classical model. Rep. Prog. Phys. **52**(8), 945 (1989)
11. Clayton, D.D., Rassbach, M.E.: Termination of the s-process. Astrophys. J. **148**, 69–85 (1967)
12. Goriely, S.: Uncertainties in the solar r-abundance distribution. Astron. Astrophys. **342**, 881–891 (1999)

13. Langanke, K., Martinez-Pinedo, G.: Nuclear weak-interaction processes in stars. Rev. Mod. Phys. **75**(3), 819–862 (2003)
14. Beer, H., Käppeler, F.: Neutron capture cross section on ^{138}Ba, $^{140,142}Ce$, $^{175,176}Lu$ and ^{181}Ta at 30 keV: prerequisite for investigation of the ^{176}Lu cosmic clock. Phys. Rev. C **21**(2), 534 (1980)
15. Doll, C., Börner, H.G., Jaag, S., Käppeler, F., Andrejtscheff, W.: Lifetime measurement in ^{176}Lu and its astrophysical consequences. Phys. Rev. C **59**(1), 492 (1999)
16. Karakas, A.I.: Nucleosynthesis of low and intermediate-mass stars. In: Goswami, A., Reddy, B.E. (eds.) Principles and Perspectives in Cosmochemistry, pp. 107–164. Springer, Heidelberg (2010)
17. Heil, M., Käppeler, F., Uberseder, E., Gallino, R., Pignatari M.: Neutron capture cross sections for the weak s process in massive stars. Phys. Rev. C **77**(1), 105808 (2008)
18. Woosley, S.E., Hoffman, R.D.: The alpha-process and the r-process. Astrophys. J. **395**, 202–239 (1992)
19. Qian, Y.-Z.: The origin of the heavy elements: recent progress in the understanding of the r-process. Prog. Part. Nucl. Phys. **50**, 153–199 (2003)
20. Kar, K.: Weak interaction rates for stellar evolution, supernovae and r-process nucleosynthesis. In: Goswami, A., Reddy, B.E. (eds.) Principles and Perspectives in Cosmochemistry, pp. 183–208. Springer, Heidelberg (2010)
21. Cameron, A.G., Cowan, J.J., Truran, J.W.: The waiting-point approximation in r-process calculations. Astrophys. Space Sci. **91**, 235–243 (1983)
22. Arnould, M., Goriely, S.: The p-process of stellar nucleosynthesis and nuclear physics status. Phys. Rep. **384**(1–2), 1–84 (2003)
23. Arnould, M.: The evolution of massive stars and the concomitant non-explosive and explosive nucleosynthesis. In: Goswami, A., Reddy, B.E. (eds.) Principles and Perspectives in Cosmochemistry, pp. 277–343. Springer, Heidelberg (2010)
24. Qian, Y.-Z., Woosley, S.E.: Nucleosynthesis in neutrino driven winds: I. The physical conditions. Astrophys. J. **471**, 331–351 (1996)
25. Horowitz, C.J.: Weak magnetism for antineutrinos in supernovae. Phys. Rev. D **65**(4), 043001 (2002)
26. Meyer, B.S., McLaughlin, G.C., Fuller, G.M.: Neutrino capture and r-process nucleosynthesis. Phys. Rev. C **58**(6), 3696 (1998)
27. Mention, G., Fechner, M., Lasserre, Th., Mueller, Th. A., Lhuiller, D., Cribier, M., Letourmeau, A.: Reactor antineutrino anomaly. Phys. Rev. D **83**(7), 073066 (2011)
28. Guinti, C., Laveder, M., Li, Y.F., Liu, Q.F., Long, H.W.: Update of short-baseline electron neutrino and antineutrino disappearance. Phys. Rev. D **86**(11), 113014 (2012)
29. Malkus, A., Kneller, J.P., McLaughlin, G.C., Surman R.: Neutrino oscillations above black hole accretion disks: disks with electrn-flavor emission. Phys. Rev. D **86**(8), 085015 (2012)
30. Chakraborty, S., Choubey, S., Goswami, S., Kar, K.: Collective flavor oscillations of supernova neutrinos and r-process nucleosynthesis. J. Cosmol. Astropart. Phys. **1006**, 007 (2010)
31. Fischer, T., Gangi, G., Dzhioev, A.A., Martinez-Pinedo, G.M.: Neutrino signal from proto-neutron star evolution: effects of opacities from charged-current-neutrino interactions and inverse neutron decay. Phys. Rev. C **101**(2), 025804 (2020)
32. Möller, P., Nix, J.R., Kratz, K.L.: Nuclear properties for astrophysical and radioactive-ion-beam applications. At. Data Nucl. Data Tables **66**(2), 131–343 (1997)
33. Kitaura, F.S., Janka, H.-T., Hillebrandt, W.: Explosions of O-Ne-Mg cores, the Crab supernova, and subluminous type II-P supernovae. Astron. Astrophys. **450**, 345–350 (2006)
34. Janka, H.-T., Müller, B., Kitaura, F.S., Buras, R.: Dynamics of shock propagation and nucleosynthesis conditions in O-Ne-Mg core supernovae. Astron. Astrophys. **485**(1), 199–208 (2008)
35. Wanajo, S., Janka, H.-T., Müller, B.: Electron capture supernovae as the origin of elements beyond iron. Astrophys. J. **726**(1), L15 (2011)

36. Mirrizi, E., Tamborra, I., Janka, H.-T., Saviano, N., Bolig, R., Hudenpohl, L., Chakraborty, S.: Supernova neutrinos: production, oscillations and detection. Riv. Nuovo Cim. **39**, 1–112 (2016)
37. Nishimura, S., Kotake, K., Hashimoto, M.-A., Yamada, S., Nishimura, N., Fujimoto, S., Sato, K.: r-process nucleosynthesis in magnetohydrodynamic jet explosions of core collapse supernovae. Astrophys. J. **642**, 410–419 (2006)
38. Lattimer, J.M., Schramm, D.N.: Black-hole-neutron-star collisions. Astrophys. J. **192**, L145–L147 (1974)
39. Li, L.-X., Paczyński, B.: Transient events from neutron star mergers. Astrophys. J. **507**(1), L59–L62 (1998)
40. Metzger, B.D., Martinez-Pinedo, G., Darbha, S., et al.: Electromagnetic counterparts of compact object mergers powered by the radioactive decay of r-process nuclei. Mon. Not. Roy. Soc. **406**(4), 2650–2662 (2010)
41. Aboussir, Y., Pearson, J.M., Dutta, A.K., Tondeur, F.: Nuclear mass formula via an approximation to the Hartree-Fock method. At. Data Nucl. Data Tables **61**, 127–176 (1995)
42. Wang, N., Liu, M., Wu X.: Modification of nuclear mass formula by considering isospin effects. Phys. Rev. C **81**(4), 044322 (2010)
43. Liu, M., Wang, N., Deng, Y., Wu, X.: Further improvements on a global nuclear mass model. Phys. Rev. C **84**(1), 014333 (2011)
44. Duflo, J., Zuker, A.P.: Microscopic mass formulas. Phys. Rev. C **52**(1), R23 (1995)
45. Goriely, S., Channel, N., Pearson, J.M.: Further exploration of Skyrme-Hartree-Fock-Bogoliubov mass formulas XVI. Inclusion of self-energy effects in pairing. Phys. Rev. C **93**(3), 034337 (2016)
46. Goriely, S., Channel, N., Pearson, J.M.: Further exploration of Skyrme-Hartree-Fock-Bogoliubov mass formulas XII. Stffness and stability of neutron star matter. Phys. Rev. C **82**(3), 035804 (2010)
47. Sun, B.-H., Meng, J.: Challenge on the astrophysical r-process calculation with nuclear mass models. Chin. Phys. Lett. **25**, 2429 (2008)
48. Audi, G., Wapstra, A.H., Thibault, C.: The AME2003 atomic mass evaluation: (II) tables, graphs and references. Nucl. Phys. A **729**(1), 337–676 (2003)
49. Wang, M., Audi, G., Watpstra, H., Kondev, F.G., MacCormick, M., Xu, X., Pfeiffer, B.: The AME2012 atomic mass evaluation. Chin. Phys. C **36**, 1603 (2012)
50. Gill, R.L., Casten, R.F., Warner, D.D., Piotrowski, A., Mach, A., Hill, J.C., Wohn, F.K., Winger, J.A., Morch, R.: The first measurement of a r-process waiting-point nucleus. Phys. Rev. Lett. **56**(17), 1874 (1986)
51. Kratz, K.-L., Gabelmann, H., Hillebrandt, W., Pfeiffer, B., Schlösser, K., Thielemann, F.-K.: The beta-decay half-life of $^{130}_{48}Cd_{82}$ and its importance for astrophysical r-process scenarios. Z. Phys. A **325**, 489–490 (1986)
52. Pfeiffer, B., Kratz, K.-L., Thielemann, F.-K., Walters, W.B.: Nuclear structure studies for the astrophysical r-process. Nucl. Phys. A **693**(1–2), 282–324 (2001)
53. Ikeda, K.L., Fujii, S., Fujita, J.I.: The (p,n) reactions and beta decays. Phys. Lett. **3**, 271 (1963)
54. Zhi, Q., Caurier, E., Cuenca-Garcia, J.J., Langanke, K., Martinez-Pinedo, G., Sieja, K.: Shell model half-lives including first forbidden contributions for r-process waiting-point nuclei. Phys. Rev. C **87**(2), 025803 (2013)
55. Hosmer, P.T., Schatz, H., Aprahamian, A., et al.: Half-life of the doubly magic r-process nucleus ^{78}Ni. Phys. Rev. C **94**(11), 112501 (2005)
56. Johnson, C.W., Koonin, S.E., Lang, G.H., Ormand, W.E.: Monte Carlo methods for the nuclear shell model. Phys. Rev. Lett. **69**(22), 3157 (1992)
57. Koonin, S.E., Dean, D.J., Langanke, K.: Shell model Monte Carlo methods. Phys. Rep. **278**(1), 1–77 (1997)
58. Ormand, W.E.: Estimating the nuclear level density with the Monte Carlo shell model. Phys. Rev. C **56**(4), R1678 (1997)
59. Langanke, K.: Shell model Monte Carlo level densities for nuclei with A ∼ 50. Phys. Lett. B **438**(3–4), 235–241 (1998)

60. Alhassid, Y., Liu, S., Nakada, H.: Spin projection in the Shell Model Monte Carlo method and the spin distribution of level densities. Phys. Rev. Lett. **99**(16), 162504 (2007)
61. Mocelj, D., Rauscher, T., Martinez-Pinedo, G., Langanke, K., Pacearescu, A., Faessler, A., Thielemann, F.-K., Alhassid, Y.: Large-scale prediction of the parity distribution in the nuclear level density and application to astrophysical reaction rates. Phys. Rev. C **75**(4), 045805 (2007)
62. Loens, H.P., Langanke, K., Martinez-Pinedo, G., Rauscher, T., Thielemann, F.-K.: Complete inclusion of parity-dependent level densities in the statistical description of astrophysical reaction rates. Phys. Lett. B **666**, 395–399 (2008)
63. Rauscher, T., Thielemann F.-K.: Astrophysical reaction rates from statistical model calculations. At. Data Nucl. Data Tables **75**(1–2), 1–351 (2000)
64. Rauscher, T., Thielemann, F.-K.: Tables of nuclear cross sections and reaction rates: an addendum to the paper "Astrophysical reaction rates from statistical model calculations". At. Data Nucl. Data Tables **79**(1), 47–64 (2001)
65. Rauscher, T.: The path to improved reaction rates for astrophysics. Int. J. Mod. Phys. E **20**(5), 1071–1167 (2011)
66. Goriely, S., Hilaire, S., Girod, M.: Latest development of the combinatorial model of nuclear level densities. J. Phys. Conf. Ser. **337**, 012027 (2012)
67. Cowan, J.J., Thielemann, F.-K., Truran, J.W.: The r-process and nucleochronology. Phys. Rep. **208**(4–6), 267–394 (1991)
68. Abbott, B. P., Abbott, R., Abbott, T. D., et al.: GW170817: observation of gravitational waves from a binary neutron star inspiral. Phys. Rev. Lett. **119**(16), 161101 (2017)
69. Abbott, B.P., Abbott, R., Abbott, T.D., et al.: Gravitational waves and gamma rays from a binary neutron star merger: GW170817. Astrophys. J. Lett. **848**(2), L13 (2017)
70. Arcavi, I.: The first hours of the GW170817 kilonova and the importance of early optical and ultraviolet observations for constraining emission models. Astrophys. J. Lett. **855**, L23 (2018)
71. Kasen, D., Metzger, B.D., Barnes, J., Quataert, E., Ramirez-Ruiz, E.: Origin of the heavy elements in binary neutron-star mergers from a gravitational-wave event. Nature **551**, 80–84 (2017)
72. Shibata, M., Fujibayashi, S., Hotokezaka, K., et al.: Modelling GW170817 based on numerical relativity and its implications. Phys. Rev. D **96**(12), 123012 (2017)
73. Baiotti, L., Rezzolla, L.: Binary neutron star mergers: a review of Einstein's richest laboratory. Rep. Prog. Phys. **80**(9), 096901 (2017)
74. Margalit, B., Metzger, B.D.: Constraining the maximum mass of neutron stars from multi-messenger observations of GW170817. Astrophys. J. Lett. **850**(2), L19 (2017)
75. Perego, A., Thielemann F.-K., Cescutti, G.: r-process nucleosynthesis from binary mergers (2021). arXiv:astro-ph/2109.09162
76. Hulse, H.A., Taylor, J.A.: Discovery of a pulsar in a binary system. Astrophys. J. Lett. **195**, L51–L53 (1975)
77. Symbalisty, E., J.M., Schramm, D.N.: Neutron star collisions and the r-process. Astrophys. J. Lett. **22**, 143–145 (1982)
78. Lattimer, J.M., Schramm, D.N.: The tidal disruption of neutron stars by black holes in close binaries. Astrophys. J. **210**, 549–567 (1976)
79. Lattimer, J.M., Mackie, F., Ravenhall, D.G., Schramm, D.N.: Decompression of cold neutron star matter. Astrophys. J. **213**, 225–233 (1977)
80. Baym, G., Bethe, H.A., Pethick, C.J.: Neutron star matter. Nucl. Phys. A **175**(2), 225–271 (1971)
81. Myers, W.D., Swiatecki, W.J.: Nuclear masses and deformations. Nucl. Phys. A **81**(1), 1–60 (1966)
82. Kulkarni, S.R.: Modelling supernova-like explosions associated with gamma-ray bursts with short durations (2005). arXiv:astro-ph/0510256
83. Tanvir, N.R., Levan, A.J., Fruchter, A.S., et al.: A 'kilonova' associated with the short-duration γ-ray burst GRB 130603B. Nature **500**, 547–549 (2013)
84. Berger, E., Fong, W., Chornock, R., et al.: A r-process kilonova associated with short-hard GRB 130603B. Astrophys. J. Lett. **774**(2), L23 (2013)

85. Metzger, B.D.: Kilonovae. Living Rev. Relat. **20**, 3 (2017)
86. Rosswog, S., Liebendorfer, M., Thielemann, F., et al. Mass ejection in neutron star mergers. Astron. Astrophys. **341**, 499–526 (1999)
87. Metzger, B.D.: Lessons from the light of a neutron star merger. Ann. Phys. **410**, 167923 (2019)
88. Cowperthwaite, P.S., Berger, E., Villar, V.A., et al.: The electromagnetic counterpart of the binary star merger LIGO/VIRGO GW170817.II.UV Optical and near infrared light curves and comparison to Kilonova models. Astrophys. J. Lett. **848**, L17 (2017)
89. Watson, D., Hansen, C.J., Selsing, J., et al.: Identification of strontium in the merger of two neutron stars. Nature **574**, 497–500 (2019)

Index

© The Author(s), under exclusive license to Springer Nature Switzerland AG 2022
D. Bandyopadhyay, K. Kar, *Supernovae, Neutron Star Physics
and Nucleosynthesis*, Astronomy and Astrophysics Library,
https://doi.org/10.1007/978-3-030-95171-9

Printed in the United States
by Baker & Taylor Publisher Services